W9-AUV-740

| DATE DUE | | | |
|---|---|---|---|
| | | | |
| | | | |
| | | | |
| | | | |
| | | | |
| | | | |
| | | | |
| | | | |
| | | | |
| | | | |
| | | | |
| | | | |

# The Two Faces of Chemistry

*Luciano Caglioti*

*Translated by Mirella Giacconi*

The MIT Press
Cambridge, Massachusetts
London, England

First MIT Press paperback edition, 1985

This book was set in Baskerville by Achorn Graphic Services, Inc. and printed and bound by Halliday Lithograph in the United States of America.

**Library of Congress Cataloging in Publication Data**

Caglioti, Luciano.
  The two faces of chemistry.

  Translation of: I due volti della chimica.
  Bibliography: p.
  Includes index.
  1. Chemical engineering—Social aspects.
I. Title.
TP149.C2513   1983      363.1'72      82-12706
ISBN 0-262-03088-8 (hardcover)
ISBN 0-262-53064-3 (paperback)

*We are like leaves*
*in the summer season that the sun nurtures*

*Mimnermus*

# Contents

# 8

# 9

# 10

# Foreword

In the early 1960s Italy, Europe, and most of the world sailed forth in a general euphoria only slightly dimmed by the clouds that seemed to be gathering over some of the recently decolonized countries. It was generally felt, in fact tacitly assumed, that with the end of the cold war between the United States and the Soviet Union, the acceptance of the nuclear balance, and the establishment of détente between the two superpowers, the sinister legacies of World War II would be settled and left behind and that the world could now look forward with confidence to a future of increasing production, increasing consumption, and increasing prosperity. The dangers of a political nature having faded, if not entirely disappeared, no others seemed to be threatening mankind except for those connected with overpopulation, which appeared remote.

Ten years later the outlook was already quite different. Various voices had been raised, some timidly and some forcefully, to warn that things could hardly go on that way forever: Continuous growth, yes, but on all fronts? And how far? Had not the time come to take stock of our global resources and to apply the brakes, if not to consumption, at least to waste, to artificially created needs, and to air, soil, and water pollution? That the time for a reassessment had indeed arrived became obvious to the whole world suddenly and painfully in the fall of 1973, during the few days of the "small" Yom Kippur War between Egypt and Israel. For the first time we confronted the fact that oil—the main source of energy for all industrialized nations and the basic raw material for thousands of by-products—far from being inexhaustible, as we had always assumed, could temporarily run short because of somebody's arbitrary decision to cut off the supply. Furthermore, a more thoughtful examination revealed that oil might run out entirely within a few decades, or more accurately, that it would definitely run out through depletion of the deposits. This sudden awareness of a deadline that could perhaps be postponed but was inevitable was quite sobering from many points of view and clearly established the need to solve intelligently and on a world scale many of the problems that had been accumulating for years. The alarm sounded: We shall run out of oil, we are already running out of it, and its disappearance will put an end to the era of cheap energy, to the *belle époque* of thoughtless waste, of gasoline aplenty and not much more expensive than mineral water. Sooner or later, moreover, we shall also run out of many minerals whose consumption has been increasing exponentially at the expense of limited resources. Thus it finally dawned on us that we have been wonderfully clever in the short run, in solving problems that, however complex, are temporary and marginal, and that at the same time we have been incredibly improvident with regard to the larger problems, the long-range, global problems on which depends nothing less than the survival of our civilization and perhaps of our species.

A new blow has thus been dealt to the Enlightenment concept of progress. A degree of caution had already entered our view of progress at the beginning of the century and particularly after World War I: One could speak of progress, yes, but only of scientific and technological progress; certainly not moral and perhaps not even cultural or artistic. Today experts as well as laymen are questioning the idea of scientific-technological progress itself. After all, the industrial revolution caused two bloody wars and chemistry produced dynamite; from Einstein and Fermi came Hiroshima, from defoliants the dioxin tragedy, from tranquilizers the thalidomide babies, and from food colorings cancer. Enough! Let us stop; let us go back.

But we cannot go back now—or rather we can, but only if we are willing to accept a loss of life of unthinkable proportions. Going back to the beginning means reopening the door to epidemics and high infant mortality. It means giving up the use of chemical fertilizers, thereby reducing agricultural production by a half or two thirds and condemning to starvation hundreds of millions of people in addition to the many who are already starving. Mankind at present finds itself in a situation so new, so critical, and so complex that it would be naive to think that we can solve our problems on the basis of one general criterion. We can neither continue to advance indiscriminately nor stop or retreat on all fronts. What we must do is to face one problem at a time with honesty, intelligence, and humility. This is the delicate but formidable task that confronts today's and tomorrow's scientists, and this is the central theme of Professor Caglioti's book.

Rather than a chemistry handbook, I would say, this is a manual of practical behavior. It is good—indeed, essential—to free the many and serious problems currently confronting us from the context of emotionalism and special interests and to clarify them with competence and sincerity. Not always, rather seldom in fact, will the author suggest a solution. An objective examination of the facts sometimes reveals that a problem does not exist; or, if a problem does exist, that a solution could only be found through excessively costly research; or even that the truth (as in the case of the "saccharin mess") may be drowned in a morass of contradictory data. The last, of course, is an extreme case owing to the fact that saccharin is a product whose risks as well as benefits are modest and ill defined. Quite different, and of more universal concern, is the problem of food additives since by now most human beings eat foods that have been processed or preserved. Some additives are useful, even necessary, for example, those that give longer and better protection against spoilage. Others, such as food colorings, have a purely commercial function, in that they satisfy false needs artificially created by habit or advertising. It would not be at all impossible for us to become accustomed to eating grey salami or colorless jams (that is, in their "natural" colors), but against such innovations, logical as they may be, "consumers' resistance has so far been very strong." Consequently, the same arsenal of propaganda weapons employed to promote artificial and sophisticated needs ought now to be mobilized against the use of useless additives. For in the more carefully thought-out ecological budget advocated by this book, uselessness is in effect equivalent to harmfulness. If an additive is not beneficial, it must be assumed to be harmful, however slight the danger may be. A case in point is the nitrates and nitrites that for centuries have been added to cured meats to heighten their color and are now suspected of causing cancer through the complex and unforeseeable changes they undergo upon consumption.

Just as delicate is the question of medicines. That any drug is potentially a poison was a fact already known to Hippocrates (as evidenced by the semantic ambivalence of the Greek word). We learn that "in the United States 3–5% of hospitalizations are due to a bad reaction to a medication." And nobody really knows what happens when a patient is given two or more drugs about whose compatibility and mutual interaction the general practitioner (or the pharmacologist for that matter) knows little or nothing. Not to mention the fact that many people have fallen into the habit of taking drugs without a doctor's presciption, simply on the basis of somebody else's experience or hearsay.

It is up to us, now, to weigh the benefits against the risks intelligently, competently, and objectively. In most cases, however, such an evaluation goes beyond the ability of the layman; and, by and large, we are all laymen. It is already fortunate if each of us can become competent in even one of the countless problems besetting the world. It is also difficult to be objective when it is considered that the press and mass media are constantly bombarding us with an ever increasing amount of information that is imprecise, distorted, incomplete, often ill understood by the reporters themselves, and almost always tainted by special interests or preconceived notions. A case in point is the question of tobacco, discussed at length in the book. Although people have generally become aware of the fact that "smoking is dangerous to one's health," it is instructive to read that in the German Federal Republic, for example, the government's revenues from tobacco amount to 9 billion marks per year, but that the social cost of smoking—that is, the cost of treating diseases caused directly or indirectly by smoking—adds up to 20 billion marks or that tobacco kills four times as many people as car accidents.

It is very hard to make a judgment on the toxic effects of the chemical elements that are present in trace quantities in our environment and in the food we eat. (They have always been there—the sea contains just about all of them—but concentrations are on the increase and new elements have been added.) We have known for a long time that arsenic and selenium are toxic, that is, harmful or lethal when absorbed in large amounts. But what is the meaning of "large"? On the other hand, the most modern and sophisticated methods of chemical analysis have allowed us to determine that in small amounts both metals are necessary, or at least useful: arsenic as a growth agent, selenium as a mercury antidote. One should add that the useful amounts vary considerably from species to species and very likely from individual to individual (as is probably the case for other elements and compounds as well). Consequently, while it would be wise to reduce their concentrations in the environment, it would be foolish to eliminate them completely. But where is the dividing line between wisdom and foolishness?

At the height of uncertainty and confusion, as the author points out, stands the question of energy. Yet this problem, which is tied in with all the other problems of our time (including the political ones), is the greatest of them all, the very foundation of our survival, before which all others should pale in significance; "Energy or extinction," warns Fred Hoyle in a book mentioned in these pages. It is also the problem we are least prepared to cope with since the most plausible solution, namely, the use of nuclear energy, is not supported by experience accumulated over decades or centuries, as is the case for other energy sources; moreover, it exceeds the confines of classical physics and chemistry and clashes against long-standing habits and troubling mental associations. To many people plutonium means Pluto, the atom means Hiroshima. Here, more than ever, both sides of the issue—benefits and risks—are distorted, misrepresented, and always tainted by the enormous economic interests at stake. There is no consensus on their objective evaluations even among the experts. Yet the problem cannot be set aside, for a shortage of energy would imply a loss of life of unthinkable proportions; nor can the solution be delegated to the next generation, which would thus be punished for our irresponsibility. To solve this problem—let me repeat it once more—we need intelligence, competence, and honesty.

From all we have said and from the many other vital issues discussed in the book one thing stands out clearly; the need, the moral imperative not to be naive, ignorant, and easily swayed. Never before has the need for education been so crucial, and never before have the schools, at least in Italy, been so ill equipped to educate us. We welcome all those, like Professor Caglioti, who propose to make up for these shortcomings. The problems are real and will be solved neither by shouting slogans nor by marches and sit-ins, but rather by concrete action and trusting in human rationality. There are no other means to this end. If we oppose a necessary and urgent decision, we should have a better alternative to propose. If we speak of "new patterns of development," we should understand what we are talking about. In a word, we must learn not to fall prey to thoughtless optimism or a catastrophe syndrome, not to deafen each other with words.

Underneath the statistics and technical data, which with good reason are plentiful, there flows through this book a silent current of wisdom, educational intent, and morality. While it does not attempt to dictate solutions, by its very character it teaches us how best to go about finding them. Everyone will find food for thought in it, and it is to be hoped that it will be selected as a textbook and widely circulated throughout our schools.

Primo Levi

# *Preface*

Science plays a twofold role in our world, one at the cultural level and the other at the technological level. Chemistry is no exception. Indeed, it is one of the broadest branches of science, if for no other reason that, when we think about it, everything is chemistry.

Just one measure of chemistry's impact on culture is the effect biochemistry has had on our understanding of human nature. The fundamental unity of the biological world is now an accepted concept; the same molecules, the same structures form the basis of life in both amoebas and elephants. Precise studies on protein sequences have enabled a computer to write a phylogenetic tree of the human species. Thus Darwin's theory of evolution, so strongly opposed by the religious community on the one hand, and by part of the scientific community on the other, is now supported by solid and unassailable biochemical evidence. The mechanisms responsible for the hereditary transmission of genes and the storing of information in chemical molecules are now clearly understood, at least in principle. Recent studies have established a correlation between significant characteristics of information redundancy in human DNA molecules and analogous properties of man's language, thereby raising fundamental and fascinating questions about the mechanism of language formation. Research into the effects of exogenous and endogenous drugs has cast serious doubts on the independence of behavior from contingent factors.

Chemistry's impact on technological progress is equally impressive. Chemistry is drugs, insecticides, fertilizers. Chemistry is textile fibers, glasses, glues, telephones. Chemistry is underwater cables and television sets. Chemistry has invaded our lives, has provided us with new foods and new materials, has replaced wood and metal with less expensive products, has enabled low-income classes to acquire things that otherwise would have been inaccessible.

But the growth of the chemical industry has left in its wake very serious problems—widespread pollution and the alarming diffusion of unsafe products. What are the benefits of chemistry and what are the risks? How many of these risks can and cannot be avoided? How much of the negative side of chemistry is due to economic profit, how much to improvidence? Is a certain amount of risk a necessary consequence of the acquisition of goods and the battle against disease? These are all very difficult questions and are all too often uttered by voices as strident as they are misinformed or, worse, biased when not wholly self-serving.

I do not hope to settle such issues here and now, for it is not in my power—nor perhaps in anybody else's power—to do so. My intent is simply to examine in the light of hard facts the general framework within which the concerned scientists operate.

In an effort not to overload the text, I have only given there general information on the scope of chemistry's impact on modern society. The reader will find supplementary material and quantitative data in the appendixes.

This book is the result of readings, discussions, exchanges of ideas, and advice from friends. In particular, I wish to thank my father, who has been an inexhaustible source of advice and information; Enrico Cernia, whose knowledge has been the source of various paragraphs; and Mimmo Misiti, with whom I have discussed almost every line.

# The Two Faces of Chemistry

# 1

## Chemistry and Modern Society

In many parts of the world the 30-year period from 1945 to 1976 saw a general improvement in the standard of living of which chemistry was perhaps the main promoter. Food became better and more plentiful; synthetic fibers made clothes available to people too poor to afford the cost of traditional materials; new drugs conquered old diseases. It is not an overstatement to say that society in the industrialized nations is based on the chemical industry.

Statistics show that in the OECD countries (Organization for Economic Cooperation and Development, whose members include, among others, the Western European nations, the United States, Japan, and Canada) chemical production in the decade 1963–1973 increased at a rate of 9.3% per year.[1,*] One important measure of the impact of chemistry's growth is given by the per capita consumption of chemical products. The average Italian, for example, consumed $77 worth of chemicals in 1963 and $247 worth in 1975. In the same period per capita consumption in the German Federal Republic rose from $130 to $424; in Switzerland, from $77 to $335. In the European OECD countries, the average increased from $77 to $334. Table 1.1 illustrates these changes.

From a more strictly industrial point of view, the value added per person employed in the chemical industry is considerable. For the year 1975 it was computed at $32,460 per year for the German Federal Republic, $28,300 for Belgium, and $21,600 for Italy. These figures, listed in table 1.2, form a basis for considering the economic aspects of the chemical industry in general and, in particular, the relative degrees of production in the individual countries. (It is worth

*Numerical superscripts indicate reference citations; alphabetical superscripts indicate notes. Notes and references are to be found at the back of the book.

**Table 1.1**
Estimated per capita consumption of chemicals in the OECD countries from 1963
(in dollars)

| Country | 1963 | 1970 | 1973 | 1974 | 1975 |
|---|---|---|---|---|---|
| Austria | 56[a] | 121 | 285 | 402 | n.a.[b] |
| Belgium | 70 | 154 | 297 | 405 | 334 |
| Denmark | n.a. | n.a. | 248 | 336 | 329 |
| Finland | n.a. | 130 | 227 | 368 | 405 |
| France | 80 | 141 | 265 | 335 | 340 |
| Italy | 77[a] | 115 | 186 | 254 | 247 |
| Norway | 73 | 134 | 208 | 285 | 325 |
| Holland | 76[a] | 172 | 212 | 338 | 294 |
| Portugal | 14[a] | 47 | 78 | n.a. | n.a. |
| United Kingdom | 113 | 165 | 246 | 347 | 343 |
| German Federal Republic | 130 | 182 | 306 | 392 | 424 |
| Spain | 39 | 96 | 196 | 258 | 281 |
| Sweden | 89 | 174 | 261 | 394 | 412 |
| Switzerland | 77[a] | 155 | 271 | 364 | 335 |
| Average OECD, Europe | 77[a] | 143 | 242 | 331 | 334 |
| Canada | 77[a] | 109 | 185 | 283 | 285 |
| Japan | 59 | 141 | 242 | 294 | 294 |
| United States | 160 | 227 | 290 | 358 | 377 |

Source: *L'Industrie chimique 1974/75* for the figures up to 1973; *L'Industrie chimique 1975*
for the 1974 and 1975 figures.
a. Excluding man-made fibers.
b. n.a. = not available.

noting that since 1973 the added value for employees in the United
States has been *twice* that of Italy—$34,870 versus $17,980 in 1973).

As far as employment is concerned, in 1975 the Italian chemical
industry provided jobs for 294,500 people, 39,000 of which were in
the fiber sector. Technical and clerical employees numbered 91,800,
versus 163,700 blue-collar workers. To these must be added 7,300
employees and 31,700 workers in the textile industry. In Italy white-
collar workers constitute 36.4% of the chemical labor force, while in
the United States and in the German Federal Republic the percentage
is higher, about 42%.

Chemistry is thus a source not only of goods but of jobs, and it is a
stimulus for related industries. The size of the principal chemical
firms is immense; their sales volumes are enormous. Table 1.3 lists the
10 largest chemical producers in the world in order of sales volumes.

Such impressive growth was undoubtedly favored by the fact that
the prices of crude oil remained practically unchanged from 1948 to

**Table 1.2**
Value added per person employed in the chemical industry in the OECD countries
from 1963 (in dollars)

| Country | 1963 | 1970 | 1973 | 1974 | 1975 |
|---|---|---|---|---|---|
| Austria | n.a.[a] | n.a. | n.a. | 14,640 | n.a. |
| Belgium | 5,040 | 11,400 | 22,600 | 31,300 | 28,300 |
| Denmark | n.a. | n.a. | n.a. | 23,560 | 26,400 |
| Finland | 6,600[b] | 9,180 | 12,850 | 20,120 | 22,390 |
| France | 7,530[b] | 11,890 | 19,280 | 23,360 | 24,600 |
| Italy | 6,110[b] | 11,020 | 17,980 | 21,730 | 21,610 |
| Norway | 6,860[b] | 11,770 | 15,250 | 19,420 | 23,420 |
| Holland | 5,400[b] | 9,710 | 22,100 | n.a. | n.a. |
| United Kingdom | 5,960 | 8,370 | 13,980 | 17,750 | 18,990 |
| Spain | 3,110 | 5,710 | 10,500 | 11,585 | 15,330 |
| Sweden | 7,350 | 13,400 | 21,950 | 29,400 | n.a. |
| Average Europe | 5,940 | 10,270 | 17,390 | 22,780 | 24,390 |
| Japan | 4,680 | 12,610 | 22,300 | 24,740 | 22,720 |
| United States | 18,500 | 23,860 | 34,870 | 42,520 | n.a. |

Source: *L'Industrie chimique 1974/75* for the figures up to 1973; *L'Industrie chimique 1975* for the 1974 and 1975 figures.
a. n.a. = not available.
b. Excluding man-made fibers.

1973 since oil is the main source of not only energy but the most important raw materials in industrial chemistry (with the exception of metals and cellulose). But following the increase in the price of crude oil, the rate of production took a steep downturn. Figure 1.1 shows that this downturn, already appreciable in 1974, reached in the first 6 months of 1975 a rate of 14% in the European countries and 21% in Japan; figure 1.2 illustrates the performance of the chemical industry in its various production branches in this period. These price increases affected the chemical industry more than manufacturing in general: In Italy, for example, from 1973 to 1974 chemical prices rose by about 59.4%, compared with a 31% increase in the price of manufactured products; for the United States, these figures are, respectively, 33.5% and 22%; for France, 31.7% and 27.7%; and for Great Britain, 28.5% and 24.7%. In Western Europe the consumption of chemical products amounted to $105 billion in 1974, compared to $81 billion in 1973; for the United States these figures are $75 billion and $61 billion, respectively. From 1973 to 1974, however, prices rose by an average of 30%; hence the increase in consumption is more apparent than real.

**Table 1.3**
The ten leading chemical producers in 1977

| Company | Country | Principal products | Sales ($\times\$10^3$) | Net income ($\times\$10^3$) |
|---|---|---|---|---|
| Hoechst | German Federal Republic | Pharmaceuticals, plastics, resins, fibers, dyestuffs, inorganic chemicals, agricultural products | 10,041,671 | 92,969 |
| E. I. du Pont de Nemours | United States | Synthetic fibers, plastics, resins, rubbers, dyes, pigments, fluorocarbons, diesel additives, industrial and agricultural chemicals, pharmaceuticals, explosives | 9,434,800 | 545,100 |
| Bayer | German Federal Republic | Dyestuffs, inorganic chemicals, pharmaceuticals, agricultural products, plastics, synthetic rubber, fibers, raw materials for polyurethane foams | 9,220,047 | 136,169 |
| BASF | German Federal Republic | Plastics, oils and gases, dyes, organic products, fertilizers, raw materials for synthetic fibers, potassium salts | 9,115,918 | 167,444 |
| Imperial Chemical Industries | United Kingdom | Dyestuffs, inorganic chemicals, explosives, plastics, fibers and textiles, paints, chemicals for agriculture and fertilizers, pharmaceuticals, metals | 8,139,127 | 394,476 |
| Union Carbide | United States | Chemicals and plastics | 7,036,100 | 385,100 |
| Dow Chemical[a] | United States | Basic organic and inorganic chemicals, plastics, metals, pharmaceuticals, agricultural chemicals, packing materials | 6,234,000 | 555,700 |
| Montedison | Italy | Organic and inorganic chemicals, fertilizers, fibers | 6,183,520 | (514,686) |
| Rhône-Poulenc | France | Chemicals, pharmaceuticals, textiles | 4,804,839 | 17,094 |
| Akzo Group | Holland | Synthetic fibers, organic and inorganic chemicals, pharmaceuticals, coating materials, chemical specialties, food products | 4,252,527 | (67,865) |

Adapted from *Fortune* 92(3):172 (1978).
a. From D. M. Kiefer, "Big chemical producers post moderate growth," *Chemical and Engineering News* 56(18):40 (1978).

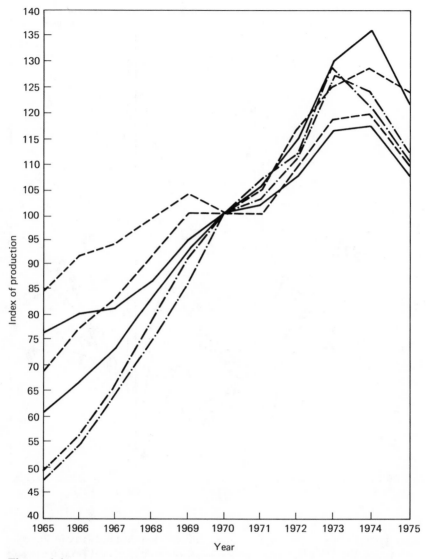

**Figure 1.1**
Index of production (1970 = 100) for the European OECD countries,
the United States, and Japan. Key to production (chemical/total industrial) by country: ——————, Europe; ——————, United States; —.—.—.,
Japan. (From *L'industrie chimique 1974/75*, modified by data from *L'industrie
chimique 1975*)

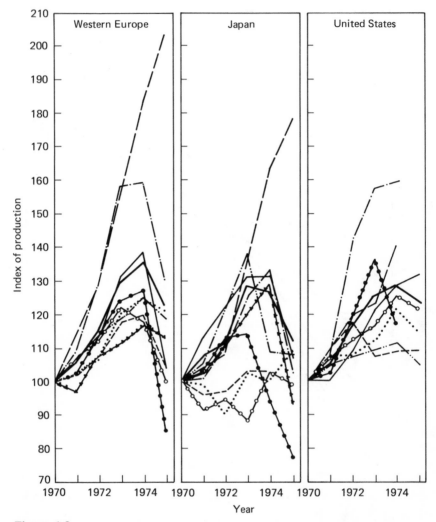

**Figure 1.2**
Production trends in the major branches of the chemical industry in Europe,
Japan, and the United States. Key: ———, organic chemical products; ---,
sulfuric acid; —·—·—·, plastics; ●●●-, dyestuffs; · · ·, nitrogenous fertilizers; -O-O-O-, phosphatic fertilizers; —··—··—··, paints and varnishes;
———, pharmaceuticals; -▲-▲-▲-, soaps and detergents; ———, total
chemical industry. (From *L'industrie chimique 1975*)

*Chemistry and the Environment*

As these figures show, the amount of chemicals produced and sold throughout the world is truly enormous. According to studies released by the EEC (European Economic Community), world production of organic compounds alone, excluding lubricating oils, amounted to 7 million tons in 1950, 63 million in 1970, and is expected to reach 250 million by 1987. A considerable fraction of these products percolates into the environment. According to fairly reliable calculations, if we take into account the rate at which these substances break down in the environment, it can be shown that in 1975 our planet was contaminated by 60–100 million tons of synthetic organic compounds. To these already fantastic amounts we must add lubricating oils, metals, solid waste, urban refuse, fuel combustion gases, sulfur compounds, oxidants, and so forth, all of which pollute our soil, water, and air.

Despite the abatement measures adopted by urban communities and industries, part of the polluting load is not eliminated and ends up in the water we drink, and in the air we breathe. The most dangerous water pollutants are toxic metals (such as lead, selenium, arsenic, chromium, and cadmium), chloride derivatives (some industrial solvents), nitrates, and detergents. Among the numerous air contaminants the most hazardous to human health are carbon monoxide, generated by furnaces and automobile engines; sulfur oxides, produced in the combustion of sulfur-containing fuels such as diešel, oil, and coal; nitrogen oxides, generated mostly by motor vehicles; residues of unburnt hydrocarbons; soot; and industrial dusts.

Once in the atmosphere, moreover, these pollutants often react with each other or with normal atmospheric constituents. Such interactions, which may be activated by ultraviolet radiation or the ozone present at high altitudes, result in the formation of a variety of compounds that are among the most troublesome to control. The most dangerous of these secondary pollutants are peroxide compounds, which are associated with the major air pollution disasters in Pennsylvania in October 1958 (20 persons died and 6,000 became ill) and in London in December 1962 (thousands of casualties, especially among elderly people suffering from chronic bronchitis).

*Toxicity of Chemicals*

With the exception of pharmaceutical products, food additives, pesticides, and herbicides, only a few of the many chemicals introduced into commerce every year undergo extensive toxicological tests. Although care is taken with notoriously unsafe products such as arsenic derivatives, lead salts, and toxic gases like carbon monoxide or phosgene, most chemicals are usually marketed and used without great precautions.

Intoxication, however, can occur, and often does, from the prolonged absorption of substances that are not known to be highly toxic but nonetheless possess a degree (albeit small) of toxicity. When absorbed by the organism over long periods of time, even in small amounts, such substances may accumulate beyond the threshold level and cause intoxication symptoms. By and large, it is extremely difficult to establish a cause-and-effect relation between poisoning and the responsible toxin. People move from place to place, eat all kinds of things, drink alcoholic beverages, take medications, breathe exhaust fumes. Frequently, intoxication is caused not by a single chemical agent but by a mixture of compounds that potentiate each other (a phenomenon known as *synergism*). It is symptomatic that the deleterious effects of protracted exposure to low-toxicity agents are usually revealed—if and when they are—by epidemiological studies conducted on factory workers who are professionally exposed to certain chemicals for long periods of time. It is in this manner that the carcinogenic properties of some substances (arsenic, asbestos, some aromatic amines, and vinyl chloride) have been discovered, as well as the toxic effects of benzene (an ingredient of glues), tri-*o*-cresyl phosphate, and other compounds used in factories as raw materials or industrial aids.

It is now generally accepted that *there is no safe substance; anything, taken in sufficient quantity, is toxic.* The toxic action may be acute and immediate or it may be a slow, insidious process. In the latter case there are two main factors that determine intoxication: accumulation of the substance in the organism and reinforcement of the effects. Many substances are retained by the organism for a length of time that varies from chemical to chemical and depends to a large extent on the conditions of the organism itself (state of health, age, metabolism, and so on). When one of these substances is absorbed with a certain continuity in either large or small amounts, if the body cannot eliminate it at a high enough rate, the substance *accumulates* and may eventually reach a level such that symptoms of intoxication develop. This is the case for compounds that are soluble in fats but not in water, such as DDT and other chlorinated hydrocarbons, organic derivatives of heavy metals (e.g., methyl mercury chloride, chief cause of the collective poisoning at Minamata, Japan), and phenol derivatives.

*Reinforcement of the effects* is a property of some carcinogenic substances. The development of symptoms in this case depends on the total amount absorbed by the organism—regardless of the length of absorption time, the size of the individual doses, and the rate of elimination. The effect persists even if the substance is eliminated from the body. Thus, if such a substance is absorbed even in minute amounts for a sufficiently long period of time, cancer may develop. This effect is sometimes due to the absorption of more than one substance, each having a certain degree of carcinogenity. The individual effects can also potentiate each other.[2]

### Chemistry and Man's Health

The multitude of chemical products to which modern man is exposed may all be toxic to some extent. Some of these health hazards arise from the voluntary absorption of chemicals, in the form of cigarette smoke, foods, alcohol, cosmetics, and other commonly used products. Others are beyond individual control. Pollution affects us in all our functions. Air contaminants with irritating effects (peroxides, ozone, and soot, for example, all of which are found in urban smog) cause eye troubles (keratitis, allergic conjunctivitis) and respiratory ailments (allergic rhinitis, laryngitis, acute and chronic bronchitis, asthma, emphysema). Two other diseases of environmental origin involving the respiratory tracts are silicosis and asbestosis, both of which, however, are associated with specific professional activities. Our skin is particularly vulnerable to radiation and atmospheric pollutants. Dyes, detergents, and even textiles can have a pathogenic effect on it. And through the skin we can absorb noxious substances such as hydrocarbons and lead compounds formed from tetraethyl lead.

In addition to specific ailments for which a cause-and-effect relation between responsible agent and pathology can be ascertained with a certain precision, man faces a very serious problem in the growing incidence of cancer. Epidemiological studies conducted in the United States, France, and Great Britain show beyond any doubt that a definite connection exists between the spread of chemical products and cancer development. According to the 1975 report of the Council on Environmental Quality,[3] from 1900 to 1970 cancer as a cause of death rose from eighth place (3.7% of the total number of deaths) to second (17% of the total); see table 1.4. (It should be noted, however, that this

**Table 1.4**
Leading causes of death: 1900, 1960, and 1970

| Rank | Cause of death | Deaths per 100,000 population | Percent of all deaths |
|---|---|---|---|
| | **1900: all causes** | (1.719) | (100) |
| 1 | Pneumonia and influenza | 202.2 | 11.8 |
| 2 | Tuberculosis (all forms) | 194.4 | 11.3 |
| 3 | Gastritis, etc. | 142.7 | 8.3 |
| 4 | Diseases of the heart | 137.4 | 8.0 |
| 5 | Vascular lesions affecting the central nervous system | 106.9 | 6.2 |
| 6 | Chronic nephritis | 81.0 | 4.7 |
| 7 | All accidents[a] | 72.3 | 4.2 |
| 8 | Malignant neoplasms (cancer) | 64.0 | 3.7 |
| 9 | Certain diseases of early infancy | 62.5 | 3.6 |
| 10 | Diphtheria | 40.3 | 2.3 |
| | Total | | 64 |
| | **1960: all causes** | (955) | (100) |
| 1 | Diseases of the heart | 366.4 | 38.7 |
| 2 | Malignant neoplasms (cancer) | 147.4 | 15.6 |
| 3 | Vascular lesions affecting the central nervous system | 107.3 | 11.3 |
| 4 | All accidents[b] | 51.9 | 5.5 |
| 5 | Certain diseases of early infancy | 37.0 | 3.9 |
| 6 | Pneumonia and influenza | 36.0 | 3.5 |
| 7 | General arteriosclerosis | 20.3 | 2.1 |
| 8 | Diabetes mellitus | 17.1 | 1.8 |
| 9 | Congenital malformations | 12.0 | 1.3 |
| 10 | Cirrhosis of the liver | 11.2 | 1.2 |
| | Total | | 85 |
| | **1970: all causes** | (945.3) | (100) |
| 1 | Diseases of heart | 362.0 | 38.3 |
| 2 | Malignant neoplasms (cancer) | 162.8 | 17.2 |
| 3 | Cerebrovascular diseases | 101.9 | 10.8 |
| 4 | Accidents | 56.4 | 6.0 |
| 5 | Influenza and pneumonia | 30.9 | 3.3 |
| 6 | Certain causes of mortality in early infancy[c] | 21.3 | 2.2 |
| 7 | Diabetes mellitus | 18.9 | 2.0 |
| 8 | Arteriosclerosis | 15.6 | 1.6 |
| 9 | Cirrhosis of the liver | 15.5 | 1.6 |
| 10 | Bronchitis, emphysema, and asthma | 15.2 | 1.6 |
| | Total | | 85 |

Source: Council on Environmental Quality, *Environmental Quality: Report VI.*
a. Violence would add 1.4%; horse, vehicle, and railroad accidents provide 0.8%
b. Violence would add 1.5%; motor vehicle accidents provide 2.3%; railroad accidents provide less than 0.1%
c. Birth injuries, asphyxia, infections of newborn, ill-defined diseases, immaturity, etc.

increase is made more impressive by the concurrent decline of other diseases, once terminal but now curable.)

There is general agreement on the estimate that 60–90% of all cancer cases can be traced to environmental factors, both natural and man-made. Preeminent among them is smoking, followed by natural radiation and then by the absorption of natural or synthetic chemicals. With regard to environmental influence on cancer development it is interesting to read the following excerpt from an article by John N. P. Davies and his coworkers:

A growing body of evidence suggests that the environment plays an essential role in the development of most human cancers. . . . Cancer registries have demonstrated major differences in occurrence of each anatomic type of cancer in various parts of the world. . . . This can be easily exemplified by briefly examining the situation with respect to gastrointestinal cancer. . . .

Cancer of the stomach is a major site of cancer death. It is the fifth leading cause of death from cancer in the United States and the leading cause of death from cancer in Italy. The frequency of stomach cancer has been declining rapidly in many countries during recent years. In the United States, for example, the rate has fallen by more than 50% in the last thirty years, and in Scandinavian countries such as Norway, an extremely high incidence area, a smaller but significant decline has been observed in recent years.

There are wide geographic variations in the occurrence of stomach cancer, it being particularly common in Japan, Iceland, Chile and several European countries. The rates in Australia, Canada and the United States are relatively low. The frequency of gastric cancer appears to be higher in Northern China than in the southern part of this country and may partially be related to the greater coarseness of food in the former. In addition, a comparison of the diets in Japan and Iceland, both areas of high incidence, demonstrated marked dietary differences. These differences both within a given country (e.g. China) and between countries (Japan and Iceland) suggest a multiplicity of environmental agents not necessarily the same.

Observations of migrating populations are of great interest. Persons moving to another country, and their children born in that country, tend to take on the cancer risk of the country to which they migrate. This is true for many types of cancer. Stomach cancer is more common in Poland than in the United States. Americans of Polish descent, however, have a rate similar to that of other Americans. Japanese living in Japan have a higher risk of developing stomach cancer than Japanese living in California. The opposite is true for prostate and colon cancers, diseases common in the United States.

Within a particular community, people of different income levels may have different cancer risks. Stomach and cervix cancers are more common among poor persons, while breast cancer, Hodgkin's disease, and perhaps prostate cancer and leukemia are more common in upper-income groups.[4]

Let us consider some specific cases. In June 1973 the EPA (Environmental Protection Agency) reported that asbestos microfibers had been found in the drinking water of Duluth, Minnesota. These fibers originate from the dumping of 67,000 tons a day of taconite, a low-grade iron ore processed by the Reserve Mining Company. As asbestos has marked carcinogenic properties, a number of measures have been taken to protect the local population from a danger whose magnitude is still unknown; the first symptoms of cancer from asbestos appear after a lag phase of about 20 or 30 years.

Another case of chemically induced cancer involves a synthetic compound, vinyl chloride. In January 1974 the National Institute of Occupational Safety and Health announced that the Goodrich Company had determined that the death by angiosarcoma of three workers involved in the production of PVC (polyvinyl chloride) could be related to the prolonged absorption of even small amounts of vinyl chloride during working hours.

Since then, additional cases of angiosarcoma have been found among workers in PVC plants and the carcinogenic properties of vinyl chloride are now clearly established. What is not known with any certainty is the number of workers that will be affected since the induction period for cancer is generally longer than 15 years and PVC production started just about 15 years ago.

Statistical studies conducted in England show that the death rate for lung cancer is far higher in cities and industrial areas than in the country: 64 in 100,000 in rural areas, 84 in towns with less than 50,000 people, and 112 in cities with more than 100,000 people. Similar results have been found by Dutch and Norwegian researchers in their respective countries.[4]

The prime culprit is tobacco. Some scientists estimate that cigarette smoking is responsible for 70% of all cases of lung cancer (The remaining 30% would be due to environmental factors). Furthermore, tobacco may act in synergism with other pollutants of the environment and with the carcinogens absorbed from the myriad of chemicals we are exposed to in our daily lives. In this light, the statement made in April 1974 by Benjamin F. Byrd, president of the American Cancer Society, appears entirely justified: "Wipe out smoking, and you eliminate some 15 to 20% of all cancer deaths."[5]

# 2

## Chemistry and Nutrition

Hunger, one of mankind's oldest traveling companions, still walks by his side throughout much of the world. The food situation is so critical that international organizations constantly solicit the help of the producing nations in creating and distributing surplus stores of cereals to Third World countries, while they work to increase agricultural production there.

From the point of view of nutrition there are two worlds. Approximately one third of the population, located mostly in Europe, North America, and Japan, lives in extraordinary luxury and is able to afford a supernutrition that often borders on waste. The remaining two thirds live more or less in poverty. Fully half of mankind suffers from hunger.

The current situation abounds in disparities and contradictions that are very hard to cure. While people are starving in Bangladesh and the Sahel area, the EEC countries have a surplus of over a million tons of powdered milk, and hundreds of thousands of tons of apples, tomatoes, and peaches are destroyed in France and Italy. (see the article by Giraudo in the 22 November 1975 issue of *Le Monde* for an explanation of the basic causes.)

Notwithstanding such discrepancies—there are countless similar situations—the level of nutrition can be said to have improved everywhere in the last three decades. The system of intensive agriculture pioneered by the United States has brought food production to levels that were unthinkable only 30 years ago; a well-coordinated international effort has saved the Third World from a catastrophe of major proportions; and the People's Republic of China has become self-sustaining with agricultural practices based on traditional models.

In the United States only 5% of the population is engaged in farming and animal husbandry. Every farmer produces enough food to feed 45 people beside himself. The cost of food constitutes 17% of the average American salary. There is a vast surplus of wheat, corn, sorghum, and soybeans. Large amounts of cereals are used as livestock feeds. Food technology is so advanced that perishable foods can be preserved, distributed, and sold months after being produced. The US agricultural model has been adopted in Canada and Europe as well.

On the other hand, in the developing countries such as India, Pakistan, and most of the African and Asian nations, 70–80% of the people are engaged in farming, which is, however, farming at the mere subsistence level. The soil is often overtaxed to the point of erosion, yields per acre are low, and a poor harvest creates immediate problems of survival for the population. Table 2.1 shows the percentage of farm labor in the total labor force in selected countries from 1960 to 1975, as well as estimates up to the year 2000.

In the 1960s a great effort was made by agencies connected with the FAO (Food and Agriculture Organization of the United Nations) to remedy such a precarious situation. Under the leadership of Norman Borlaug, who was awarded the 1970 Nobel Peace prize for his work, a

**Table 2.1**
Percentage of farm labor force in world labor force 1960–2000

| Area | 1960 | 1965 | 1970 | 1975 | 1980 | 1990 | 2000 |
|---|---|---|---|---|---|---|---|
| Developed market economies | 20.2 | 16.5 | 12.9 | 10.4 | 8.4 | 5.5 | 3.8 |
| North America | 7.1 | 5.6 | 4.1 | 3.1 | 2.4 | 1.6 | 1.2 |
| Western Europe | 23.5 | 19.3 | 15.4 | 12.7 | 10.4 | 6.8 | 4.5 |
| Oceania | 12.0 | 10.2 | 8.7 | 7.5 | 6.4 | 4.6 | 3.4 |
| Other developed market economies | 32.6 | 26.7 | 20.9 | 16.7 | 13.4 | 9.0 | 6.7 |
| Eastern Europe and USSR | 42.2 | 35.6 | 28.7 | 23.9 | 19.8 | 13.8 | 9.6 |
| Total developed countries | 28.3 | 23.4 | 18.5 | 15.2 | 12.4 | 8.4 | 5.8 |
| Developing market economies | 70.9 | 68.0 | 65.2 | 62.0 | 58.5 | 51.2 | 44.1 |
| Africa | 80.5 | 78.0 | 75.5 | 72.4 | 69.1 | 61.8 | 53.9 |
| Latin America | 47.7 | 44.2 | 40.8 | 37.3 | 33.9 | 27.5 | 21.8 |
| Near East | 69.5 | 65.7 | 62.0 | 58.1 | 54.2 | 46.5 | 39.2 |
| Far East | 73.5 | 70.9 | 68.3 | 65.4 | 62.1 | 55.0 | 47.9 |
| Other developing market economies | 81.0 | 78.7 | 76.4 | 73.9 | 71.3 | 66.4 | 61.3 |
| Asian centrally planned countries | 75.0 | 71.5 | 68.0 | 64.2 | 60.2 | 51.3 | 41.2 |
| Total developing countries | 72.6 | 69.4 | 66.3 | 62.8 | 59.1 | 51.3 | 43.1 |
| World total | 57.7 | 54.2 | 51.0 | 47.9 | 44.9 | 39.3 | 33.8 |

Source: FAO, *La quatrième enquête mondiale de la FAO sur l'alimentation*, Rome (1977), p. 95.

series of measures were taken that resulted in what is now known as the *Green Revolution*. This expression aptly describes the spectacular increase in productivity that has taken place in Southeast Asia. In several densely populated countries where hunger was once endemic, the production of wheat, rice, and corn has risen to the point of satisfying the basic nutritional needs of the populations. Lately, however, the Green Revolution has suffered some setbacks, owing to various difficulties of a technical, political, and economic nature.

Underlying both the strong growth of American agriculture and the success of the Green Revolution are similar technical reasons: extensive use of fertilizers, phytopharmacological agents, and growth regulators; genetic selection of high-yield cereal strains and livestock; and large-scale use of farm machinery.

*Fertilizers*

"Chemical pesticides and fertilizers—coupled with farm mechanization and plant genetics—have so increased productivity that about the same amount of land harvested to supply 92 million Americans in 1910 was able to take care of the needs of 214 millions last year, with enough surplus left over to permit increasing exports of our farm products overseas."[1] It is estimated that 30–40% of the increase in productivity registered in the United States, and as much as 50% in the developing countries, can be attributed to the more extensive use of fertilizers. With reference to the US agricultural model, which has inspired European practices, statistics show that while a century ago American farmers used 320,000 tons of fertilizers per year, fertilizer consumption 50 years ago had risen to 7.2 million tons per year and at present is over 40 million tons, at a cost of more than $2 billion.

The American population has grown 5-fold since 1870 and 2-fold since 1920, whereas fertilizer production has increased 125-fold since 1870 and 6-fold since 1920. It may be interesting to see how much cereal and legume yields per acre have increased as a result of the growing use of fertilizers. As shown in figure 2.1, cereal yields per acre have risen substantially over the years, while the increase has not been at all comparable in the case of legumes. This is due to the fact that while cereals need ammonium nitrate, legumes can supply their own needs of nitrogen by a process of nitrogen fixation, that is, the conversion of atmospheric nitrogen into nitrates by special microorganisms living in their roots.[2] To reiterate this point I will report statistics from another source. Table 2.2 shows the yields in pounds per acre of selected crops in the United States, Brazil, and Argentina.

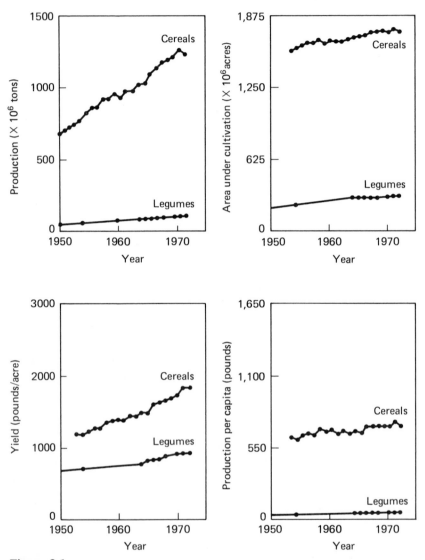

**Figure 2.1**
Comparison of trends in world production of cereal grains and grain
legumes, area under cultivation, yield, and production per capita. [From
Philip A. Abelson, *Food: Politics, Economics, Nutrition and Research,* Washington,
DC (1976)]

**Table 2.2**
Index of agricultural productivity (in pounds per acre) for the United States, Brazil, and Argentina

| Culture | United States | Brazil | Argentina |
| --- | --- | --- | --- |
| Rice | 4,428 | 1,298 | 3,400 |
| Potatoes | 21,736 | 6,248 | 9,769 |
| Sugar cane | 82,025 | 40,269 | 44,159 |
| Millet | 4,414 | 1,220 | 1,955 |
| Soybeans | 1,575 | 1,003 | – |
| Wheat | 1,776 | 813 | 1,102 |

Source: *Brasil em dados 75*, Rio de Janeiro (1975), p. 132.

The difference between Brazil's nonindustrialized agriculture and the highly industrialized agriculture of the United States is fairly dramatic, while Argentina appears to be at an intermediate stage. The difference, however, is much less marked in the case of soybeans, a legume, which confirms the fact that the application of fertilizers to legumes does not result in a substantial increase in productivity. In this case, the difference in yields between Brazil and the United States (1,000 versus 1,580 pounds per acre) can be ascribed to the use of pesticides and more modern machinery.

The consumption of fertilizers in Europe is very large, although it varies widely from country to country. For the 1973–1974 harvest, consumption of fertilizing agents (nitrogen, phosphorus, potassium) in pounds per acre was as follows: Germany, 445; Holland, 416; Belgium, 294; France, 158; Italy, 92; Great Britain, 87.[3] In 1976 sales of fertilizers in Italy amounted to $540 million.

The production and consumption of fertilizers are enormous. To meet the demand, which is constantly growing, we need nitrogen, phosphorus, potassium, hydrogen, sulfur, and above all energy. As far as phosphorus is concerned, there are enough natural deposits of phosphate rocks in the world to last for centuries. The largest deposits are in Morocco (50% of the total), the United States (30%), and Russia (15%). Additional deposits are located in North Africa (Tunisia), South Africa, the Middle East, South America, and in various islands of the Pacific and Indian oceans. By a quirk of fate, Asia's densely populated countries lack phosphate ores. The production of phosphate fertilizers, however, requires large amounts of sulfuric acid (which is also needed to make ammonium sulfate), and we must be less optimistic about the long-range prospects for the supply of sulfuric acid since the deposits of basic sulfur, which are the cheapest source of raw material for its production, are slowly being depleted.

We shall still be able to produce sulfuric acid from pyrites or natural gas containing hydrogen sulfide, but at a higher cost. There is no problem, of course, with the supply of nitrogen. To be converted into ammonia, however, nitrogen must be treated with hydrogen, which in turn originates from petrochemicals (natural gas and petroleum) and from coal (through 'water gas'). Needless to say, these commodities are running short and, in any case, can only be obtained at a much higher cost than in the past. The outlook for the future of fertilizers is therefore somewhat uncertain.

The dependence of the fertilizer industry on oil—and in a more general way on energy—has already created very serious problems for the agriculture of many countries. The increase in the price of crude oil caused a decline in fertilizer production just when demand was on the rise, owing partly to the growing nutritional needs of the affluent nations and partly to the large amounts of fertilizers needed to implement the so-called Green Revolution.

Scarcity and high prices have resulted in a decline in food production in Bengladesh, India, and all those areas where the high-yield cereal strains selected by Borlaug and others, which require extensive use of chemical fertilizers, have been adopted and cultivated as a weapon against hunger. It is a tragedy of our times that the richer nations with their greater purchasing power should compete with the poorer ones on the world's markets for both fertilizers and goods that are destined in large measure to become feed for cattle. At present, 86% of the output of fertilizers is absorbed by high-technology nations with 39% of the world's population.

In addition to the obvious benefits that world agriculture derives from the use of fertilizers, there are others not so readily perceived. The constant pressure for greater food production—and hence for greater exploitation of the soil—has often had harmful effects on arable land. Soil erosion and lowered fertility are just two of the consequences of intensive land use. The judicious use of chemical fertilizers alleviates these problems by restoring essential mineral salts and protecting the soil from erosion—a common occurrence where such a practice is not strictly observed. (However, some scientists argue that fertilizers actually promote soil erosion.) Furthermore, synthetic fertilizers are a concentrate of nutrients. Traditional fertilizers, such as manures, contain in an equal weight only about 10% of the nutrients found in the synthetic ones. Obviously, then, transporting large quantities of fertilizer favors synthetics.

One of the negative consequences of the use of fertilizers is the eutrophication of the bodies of water that receive the excess of nutrients washed out from the land. This problem has been studied, and research in the correct use of fertilizers from the environmental point of view is already showing some results. An additional and more insidious problem is the diffusion of a certain amount of heavy metals, which are essentially foreign to biological processes but are nevertheless present in chemical fertilizers as impurities. A recent report[3] warns that significant amounts of nitrous oxide ($N_2O$) formed from fertilizers added to the soil may reach the stratosphere and damage the ozone layer through a process similar to that theorized for fluorocarbons. For the time being this eventuality concerns almost exclusively the experts in the field, but it is worth mentioning in view of its potential seriousness.

*Pesticides*

In addition to fertilizers, which by themselves account for at least 40% of the increase in productivity registered in the last 40 years, chemistry provides agriculture with various other aids. Chief among them are the *pesticides* used to combat the parasites of farm crops—insects first of all, then weeds, worms, microorganisms, rodents, and so on.

Insects have always been the scourge of crops (and man—see appendix C on diseases caused by insects and on insecticides), and their devastations have been recorded throughout history. Recall the biblical swarms of locusts that left in their path hunger and famine and Homer's reference to "fumigations against insects." Coming back to the present, from 1951 to 1960 insects caused losses reckoned at about \$3.8 billion per year in the United States alone. In 1969, because of the damage caused by insects to crops and livestock, the state of California lost \$160 million, in addition to the \$125 million spent for insect control.

It is estimated that at present 30% of the potential yearly production of crops, livestock, and timber is lost to the ravages of insects and other harmful organisms. A variety of pests attacks crop plants at every stage of their development and damages foodstuff in silos, during transportation, in markets, and in the home. Scientists have identified 150 bacteria, 250 viruses, and 8,000 fungi that cause plant diseases. At least 10,000 insect species are considered destructive. There are 2,000 types of weeds in the United States and Canada alone that are responsible for consistent economic losses. Rodents, birds, and various other organisms also attack crops and other sources of food.

According to a comprehensive study,[4] in the developing countries of the Near East grain losses due to insect damage prior to harvest are estimated at around 23%. In India from 1963 to 1964 insects and rodents caused the loss of 13 million tons of wheat, or the amount needed to provide 77 million families with a loaf of bread a day for a year. In the Sudan birds devour 3,000 tons of wheat a day, and in 1971 they consumed 7 million bushels of wheat in the wheat-growing areas of the United States. In 1973 couch grass damaged 19% of grain crops in the Dakotas, and in the same year noxious grasses associated with wheat culture cost the United States and Canada $1.5 billion in direct losses and weed control.

Pests can destroy a whole industry. When the boll weevil, a cotton parasite, invaded the southeastern part of the United States, cotton production at Sea Island came to a stop. Cotton production around Tampico, Mexico, has also ceased because the cotton worm has developed a marked resistance to all available insecticides. Similarly, resistant strains of insects are developing in Texas, Louisiana, Mississippi, and Arkansas. Rodents and termites are another source of damage, particularly to poles, railroad ties, and wooden structures. This type of damage can be prevented by pretreating lumber with special preparations.

Pest control is effected on a large scale by means of insecticides, herbicides, fungicides, and so forth. The year 1973 was as critical for pesticides as it was for fertilizers, and once again the Third World countries were hardest hit by the world's crisis. From 1973 to 1974 the demand for pesticides rose by about 25%, while production decreased slightly. In part, this was a result of the general decline in productivity that followed the increase in oil prices. The FAO's *1974 Yearbook* expected the 20% shortage for the 1974–1975 period to have serious consequences for Third World nations. To quote the FAO report, "The medium-term outlook is also most disquieting, as the pesticide industry has recently been under public pressure because of the environmental effects of pesticides. Legislative obstacles have greatly increased the amount of time and money needed to develop new products."

The FAO report reflects quite accurately the Third World's concern. In the United States, on the other hand, things appear in a different light. A 1975 report on the production and sale of pesticides [5] triumphantly announced that US sales would exceed $2.3 billion by 1980, compared to $1.7 billion in 1974. The greatest expansion was expected in the herbicide and insecticide sectors; production increases were expected to benefit most wheat, cotton, and soybean crops, while fungicides were to take third place in order of production after herbicides and insecticides.

*Insecticides*

To prevent the damage caused by insects, the chemical industry has studied and developed a number of insecticides that have successfully replaced traditional compounds based on pyrethrum extracts. The era of modern insecticides began in 1939, when Paul Müller, a Swiss chemist, discovered the insecticidal properties of dichlorodiphenyl-trichloroethane (DDT), first synthesized by O. Zeidler, an Austrian student, in 1874. DDT has been of immense benefit to human health, particularly for its effectiveness in controlling malaria and other insect-borne diseases. In India, thanks to the use of DDT, malaria cases went from 75 million in 1952 to 100,000 in 1964; in Russia the number of cases dropped from 35 million in 1946 to 13,000 in 1956. According to WHO (World Health Organization), in the first 8 years alone DDT prevented at least 100 million cases of disease and 5 million deaths. In addition, the agricultural use of DDT has saved millions of tons of cereals from destruction by insects.

Beside being the first synthetic insecticide used on a world scale, DDT was the progenitor of a series of organochlorine insecticides, which includes lindane, chlordane, heptachlor, aldrin, and dieldrin, all developed and used after DDT (see appendix B). Chlorinated hydrocarbons are generally very stable compounds. Since they do not break down, or break down very slowly, they may remain in the environment for months and even years. (I will return shortly to the question of persistence, which constitutes one of the most serious drawbacks of this type of insecticide.) Organophosphorous compounds are a second group of insecticides extensively used in agriculture. Best known in this series are Parathion, Malathion, and some of their derivatives. Organophosphates are highly toxic compounds, but most decompose fairly rapidly in the environment. Analogous properties (high toxicity and easy degradability) characterize a third and more recent series of insecticides, the carbamate group, the most important of which, from a commercial point of view, is Sevin.

Insecticides, and chlorinated hydrocarbons in particular, were the starting point of the ecological campaign. More precisely, the insecticide controversy awakened the public to the need for a careful ᵛvaluation of the dangers associated with a given technology, however beneficial it may be in other respects. To be effective, an insecticide must remain in the soil for the time needed to perform its function. Long-lasting insecticides are therefore particularly well suited to the task. Their very persistence, however, represents a serious health hazard. These substances tend to accumulate in lower organisms and as they are passed up the food chain become increasingly concentrated. Thus both higher animals and man are liable to absorb relatively large amounts of toxic substances. Figure 2.2 is a schematic representation of the process of bioaccumulation of DDT.

Another serious drawback to the use of synthetic insecticides is the development of genetic resistance. The action of a poison tends to select out those members of an insect species that are not affected by it. Since insects are capable of rapid evolutionary changes, succeeding generations may have genetic traits that make them less susceptible to a given poison. This process, repeated over years and with different insecticides, can result in the spread of insect populations resistant to most chemical agents. Genetic selection can create a far more serious problem than the one we currently face since we may find ourselves weaponless against it.

As a consequence of the debates sparked by public awareness of the detrimental effects of persistent insecticides, most chlorinated hydrocarbons were banned. The use of DDT was severely restricted by the EPA (Environmental Protection Agency) in 1972, and its sale is now forbidden in the United States. In 1973 the use of aldrin and dieldrin was restricted to termite control and other nondispersive applications. Chlordane and heptachlor were banned in 1975.

The ban of DDT provoked mixed reactions, partly because the legal action aimed at proscribing some insecticides was an end in itself divorced from any consideration of the availability of valid alternatives (which in fact did not exist) and any policy for farming without them. From the uproar that followed I have chosen a few statements that implicitly reflect the two extreme points of view—that of people who only see the benefits of a chemical product and that of people who only see its potential dangers.

In a lecture delivered in 1971, Norman Borlaug, promoter of the Green Revolution, deplored the campaign waged by the press against insecticides with these words: "Should agriculture be deprived of chemicals (fertilizers and pesticides) because of an aberrant legislation advocated by a powerful pressure group run by environmental maniacs who terrorize the world by predicting that it will die of poisoning, then the world will surely die, not of poisoning, but of hunger." Referring to the book *Silent Spring* [6] by Rachel Carson, Borlaug said, "The odious and hysterical campaign currently waged by panic-sowing, irresponsible environmentalists against agrochemical products originates from a best-seller. . . . The author does not mention DDT as one of the means of defending crops. . . . Worse yet, she does not mention DDT's great contribution to the fight against malaria. . ." And he added, "DDT has shown a remarkable safety record. In the

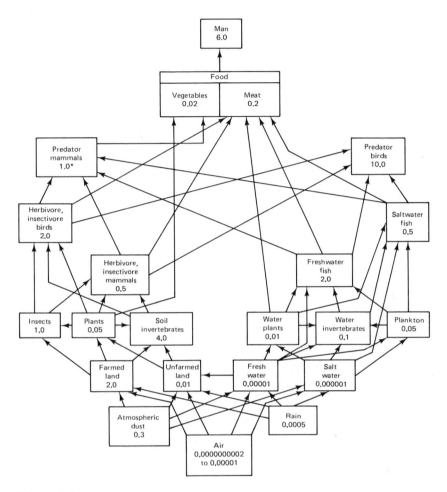

**Figure 2.2**
Concentration of DDT in the environmenta in ppm (parts per million). Asterisk indicates a figure that is uncertain due to the scarcity of data.

period of maximum production, 400,000 tons per year were used in agriculture. . . . Although hundreds of millions of people were in contact with the product in a protracted way and many were exposed to it for professional reasons, the only cases of damage to human health were the result of accidents or ingestion for suicide. There is no proof that DDT causes cancer or genetic changes in man." With regard to the battle waged by WHO against malaria, Borlaug said, "When the campaign started, around the middle of 1950, there were more than 2 million cases of malaria in Ceylon. In 1962 they had fallen to 31, in 1963 to 17. At that time the campaign of disinfestation was halted for budgetary reasons. In 1967 the number of cases had climbed back to 3,000, in 1968 to more than 16,000. . . . At the end of 1969 there were 2 million cases." The resurgence of malaria, in sum, is attributed by Borlaug to the ban of DDT.

Ivan Illich is also waging a battle—the battle against the ills of progress. *Deschooling Society*[7] is one of his works that express this concern; *Limits to Medicine,*[8] another such work, is an attack on medicine's excesses. In a lecture devoted to more general topics, Illich states that malaria, after subsiding for a while thanks to the use of chemicals, is now making a full comeback in the countries of the Third World. In his opinion, this is due in part to urbanization and the great expansion in commercial traffic, but in part also to the genetic selection of resistant strains of insects. In other words, what Borlaug attributes to the ban of DDT, Illich ascribes to the emergence of resistant strains brought about by the use of DDT.

We cannot but respect both men—their integrity is beyond question. It is not their spirit that differs, but the perspective from which they view the world's problems. Each man strives to curb an excess. Borlaug is trying to combat the extremes of poverty and hunger, and sees technology as a means of improving man's condition. Illich, on the other hand, wants to limit the excesses of consumerism, lest man become a victim of technology. Borlaug's writings and Illich's essays make instructive reading for all those people who are convinced that their own views on the matter constitute the whole truth.

The pesticide controversy has given great impetus to the search for viable alternatives to long-lasting insecticides and for new methods of pest control. Ultimately, the goal can only be achieved through an integrated system of control.

In the United States—a country where agriculture is highly industrialized and on whose exports many nations depend to meet their food requirements—pest management is a multifaceted, complex operation. Great emphasis is placed on preventive measures, which include the use of seeds free from disease or possible contamination from weeds, insects, and their larvae, and increasingly stricter custom regulations. In 1974, for example, 800,000 items (foreign seeds, insects, microorganisms, infected plants, and so forth) were confiscated from the 60 million pieces of luggage examined by custom officials. One particular episode, still remembered with horror, involves an American child who in 1967 smuggled into Miami two giant African snails. They reproduced rapidly and soon gave rise to an extremely voracious population that proceeded to destroy plantings over a large area of the city. It took half a million dollars to eradicate that pest, and damages were reckoned in the millions of dollars.

The need to reconcile the use of insecticides with the proper management of the environment has greatly stimulated chemical research in insect control. There are essentially three main fields of endeavor:

1. the search for traditional insecticides of equal effectiveness but limited side effects;
2. improvement in application techniques;
3. the study of insecticides based on entirely new concepts.

Thanks to the efforts of many scientists, progress has been made in all three fields. Insecticides have been developed that are environmentally safer than the compounds used in the last 10 years; new mixtures have been tried, as well as new application techniques (e.g., enclosing the insecticide in a porous capsule that regulates its release); finally, great strides have been made in the preparation of chemical agents that represent a totally new approach to battling insects.

### The War against the Insects

*The problems connected with the use of synthetic insecticides, the need for new weapons, and, above all, more refined techniques of chemical analysis—gas chromatography, mass spectroscopy, fractionation, and, more recently, liquid chromatography—have greatly stimulated research in the biochemistry of insects. In particular, this effort has resulted in the development of new and potentially very effective means of reducing insect populations.*

*Two groups of chemicals appear particularly promising: compounds with* morphogenetic *action (ecdysonic and juvenile hormones) and compounds that mediate* premating *phases (pheromones). Ecdysonics are hormones that regulate the molting process in insects. They are named after the first compound of the class, α-ecdysone, isolated by A. F. J. Butenandt in 1954 from 1,100 pounds of a lepidopteran,* Bombix mori. *Butenandt could only isolate a very small amount of ecdysone, and more than 10 years went by before the structure of this compound was elucidated by R. Huber and W. Hoppe. Subsequently, and with greater rapidity, similar compounds were isolated and characterized, including β-ecdysone and other derivatives.*

*Ecdysonics are found in many insect and plant species. It is not quite clear as yet what role these hormones play in plants and whether they have any relation to defense mechanisms against insects. In insects, the function of ecdysonics is to initiate the processes of* molting *(or renewing of the larva's exoskeleton) and* pupation. *When an insect is given a certain amount of synthetic ecdysonics, either molting or oogenesis and embryogenesis are profoundly altered. This results in a severe disruption in the insect's development and/or reproduction; hence the potential application of ecdysonics to insect control.*

Juvenile hormones *also play a determinant role in insect metamorphosis. The first juvenile hormone to be isolated in the laboratory was a compound extracted by C. M. Williams from a lepidopteran,* Hyalophora cecropia. *The structure of this hormone (termed JH I) was analyzed 11 years later (1967) by H. Röller. Following this pioneering research, in short order two other hormones were isolated and characterized, JH II (K. Meyer, 1968) and JH III (K. H. Trautman, 1974).*

*Juvenile hormones perform three basic functions. The first is to stimulate the larva's growth (hence the term juvenile). The second is to induce ovaric development in females and to stimulate the activity of genital glands in adults of both sexes. Lastly, they regulate the production and secretion of ecdysonics, thus orchestrating the insect's development*

*All three functions are delicate and important. Any deviation from the optimal amount of hormones can understandably cause serious disruptions in the insect's normal development. On the basis of this principle, attempts have been made to use synthetic juvenile hormones as insecticides. In 1975 one of these compounds, Altosid SR 10, was approved for commercial use in the United States and has since proved remarkably effective in the control of mosquitoes resistant to chemical insecticides.*

Pheromones *(or pherhormones, a term introduced by P. Karlson and M. Luscher in 1968 and preferred by some) are substances secreted externally that provoke a number of responses among insects of the same species, such as aggregation, alarm, trail marking, and sexual attraction. The sexual pheromones, which regulate premating behavior, are the most widely studied. They generally have a relatively simple chemical structure and are often easy to synthesize.*

*Much of the success in this field is again due to the pioneering work of Butenandt. In 1959 he succeeded in isolating 0.0004 ounces (12 milligrams) of the substance from 500,000 female silkworms and in determining its structure at a time when modern spectroscopic techniques were not yet available. His work generated an enormous amount of research that has resulted in the isolation and synthesis of a large number of hormones.*

*Pheromones are secreted by insects of one sex and sensed by individuals of the opposite sex. They are veritable sexual attractants whose function it is to guide individuals of the opposite sex to the insect that is secreting the hormone (sending the message). This function is preliminary to and essential for mating. In general, pheromones are highly specific and active at very low concentrations.*

*These substances can be applied to insect control in different ways and with different objectives. They can be used to "count" the insects of a harmful species and to determine the time of greatest infestation. This is done by placing the pheromone in special traps and counting the number of individuals captured every day. If the purpose is to combat the insects, pheromones are particularly well suited to disrupting reproduction or to mass capture. To inhibit reproduction, insects of either sex of a given species are attracted with the proper pheromone and then sterilized by radiation or chemical means. Alternatively, large amounts of pheromone can be spread over the area to be protected so that the insect, disoriented by the stimulus, which is being received from all directions, will not be able to find a mate. Mass capture is achieved by using the pheromone to concentrate insects of a given sex in a selected area, which is then sprayed with traditional insecticides.*

Insect control based on the biological criteria discussed shows great promise. The advantages are obvious: A specific insect population can be attacked at the most favorable time with products that affect only one, or a few, insect species and are safe for higher animals and man. Hormones are easily degradable, do not accumulate in the environment, and have no adverse side effects. In addition, they offer the possibility of combating insect strains resistant to chemical insecticides.[9]

However, the optimism engendered by the initial success of this technique has given way to a more cautious attitude. Great difficulties still lie ahead and further research is needed before concrete, large-scale results can be obtained. At present, therefore, insect control still depends on traditional insecticides, although they are used in a more rational manner and with greater attention to ecological effects than in the past. But we do see the possibility, in a not too distant future, of using safer, less permanent, and more selective insecticides.

One particular aspect of pest control concerns the protection of foodstuffs. A substantial fraction of stored goods is destroyed by rodents, insects, and microorganisms. The FAO estimates that as much as 10% of cereal stocks is lost every year during storage. Losses are higher in the tropical and subtropical regions than in the temperate zones. In Latin America, for example, 25–50% of cereal and legume crops is destroyed after the harvest. In Africa about a third of all agricultural produce is lost, and as much as a half in South East Asia.

Losses occur at various stages of the route that takes foodstuff from the farmer to the consumer's table: right after the harvest (in the farm-yard, so to speak), in silos, during transportation, at the mill, and, lastly, in the store. Infestations by microorganisms cause just as much damage as rodents and insects. They are most prevalent in humid, tropical or subtropical climates. The most disparate techniques are currently used for the protection of stored goods, but they are all characterized by a certain impermanence and improvisation despite the many efforts made by FAO agencies to put sound practices and modern tools at the disposal of farmers, particularly in the Third World. Some of the most common techniques include anticoagulants against rodents, fumigations with chemical agents against insects and microorganisms, elevated and underground granaries, and sackcloth packaging. None of these techniques is completely successful, and some produce adverse effects. For example, insect and rat poisons have been known to contaminate the foodstuff itself, with unpleasant consequences for the consumer.

Scientific research has recently found a way to obviate part of the problem with the development of silos in inert atmosphere, that is, silos in which the air has been replaced with nitrogen or carbon dioxide. Grains can thus be stored in an atmosphere in which rodents and insects cannot live. This practice, which is taking hold in several countries, has manifest advantages, both economic and ecological, over traditional methods.[10]

*Herbicides*
The first application of herbicides dates back to 1945 and marked a turning point in agricultural practices, as well as in the management of forests, parklands, and pastures. The first weed killers were chlorophenol derivatives. These were followed by numerous other products that are currently being sprayed on cultivated fields the world over by a variety of techniques. From 1965 to 1969 herbicide production in the United States rose at an annual rate of 17%. In 1971 sales of weed killers in the United States alone amounted to $640 million; the world figure is $1,150 million.[11]

The most common herbicides are *halogenated derivatives of phenol,* such as 2,4-D [(2,4-dichlorophenoxy) acetic acid] and 2,4,5-T [(2,4,5-trichlorophenoxy) acetic acid]; *carbamates* and *thiocarbamates; urea de-rivatives,* one of which, diuron, is extensively used in US cotton plantations; *triazine compounds* (e.g.; simazine); and *dinitroaniline derivatives,* such as trifluoralin, nitralin, and benefin.

The properties required from a weed killer are safety for animals and easy degradability in the environment. Application techniques vary widely. They range from spraying the lower part of tree bark if the purpose is to eliminate a high-trunk plant, to aerial spraying for the removal of underbrush and grasses. The benefits that agriculture derives from the use of herbicides are numerous and may be summarized as follows: less need for fertilizers since crop plants suffer less competition from weeds; greater economy in the labor needed for disinfestation practices; and a more satisfactory utilization of the land.

The disadvantages are also considerable, however. Aerial spraying, for example, is widely used in forest management to free valuable high-trunk trees like conifers from encroaching weeds, shrubs, and underbrush. After the area has been treated, rain falls directly to the ground without being in any way stopped or slowed down by shrubs and grasses, thereby increasing surface runoff and soil erosion. In addition, many of the herbicides currently marketed can have harmful effects on animals and man owing to their intrinsic toxicity, long residual action, potential accumulation or, as shown by the dioxin tragedy (Seveso, 10 July 1976), to the extremely high toxicity of some impurities.[12]

The Seveso Incident

*The Seveso incident dramatically focused world attention on the dangers connected with chemical processes and with the spread of highly toxic substances contained as impurities in some widely used products such as weed killers, insecticides, and disinfectants. The mechanics of the accident are known and there is general agreement on its reconstruction, even though the causes are still being investigated by several boards of inquiry. Following an increase in pressure within a reaction chamber during the production of trichlorophenol, the safety cap of a valve gave way and a "cloud" containing trichlorophenol, ethylene glycol, sodium carbonate, and dioxin, the last a compound of lethal toxicity [its $LD_{50}$—average lethal dose—is 0.000000001 ounces per pound (0.6 micrograms per kilogram) for guinea pigs and 0.0018 ounces per pound (115 milligrams per kilogram) for rabbits] flowed from the chamber directly into the open.*

*The amount of dioxin released in the course of this incident was exceptionally high, but it should be noted that a small amount of dioxin is normally present in trichlorophenol. Given the extreme toxicity of the substance, even a small amount is far from negligible from the point of view of the hazards it introduces into trichlorophenol and its derivatives.*

*Two products contain trichlorophenol: 2,4,5-T, a weed killer based on trichlorophenoxyacetic acid; and hexachlorophene, a bactericide (see chapter 4). The compound 2,4,5-T was used for many years as an agricultural herbicide in industrialized countries. It is also one of the components of Agent Orange, a defoliant extensively used in Vietnam by the US Army. Partly as a result of studies on the agricultural use of 2,4,5-T, partly because of the serious concern expressed by doctors and national and international agencies over the effects of Agent Orange in Vietnam, it became evident that dioxin impurities, although chemically hardly noticeable, gave both the weed killer and Agent Orange very dangerous attributes. Following the bitter controversies that raged in the United States among the public and within the Senate, the use of Agent Orange was discontinued. It has been estimated, however, that at least 1,000 pounds of dioxin must have been released on Vietnam during the war. The agricultural use of 2,4,5-T has also been forbidden or severely curtailed everywhere.*

*There is an interesting footnote to this story. In 1977 the US Army decided to destroy the supplies of Agent Orange left over from operations in Vietnam, valued at $15 million, because they contained dioxin in quantities of about 40 ppm (parts per million). In spite of extensive studies, no treatment could be found that would remove dioxin from the defoliant.*

*The Seveso incident prompted the EEC to initiate studies on a directive aimed at preventing "the risk of incidents." The purpose of the directive— familiarly known in the Community as "the Seveso directive"—is the establishment of general criteria for the regulation of hazardous industrial activities in the member countries.*

*Food Processing*

The industrialization of agriculture in the developed countries has freed a large fraction of the population from farm work. This has been made possible partly by increased productivity, partly by the development of technologies by which foods may be preserved and stored for future use. Since foods in their natural state remain edible only for very short periods of time, they must be promptly subjected to processes of stabilization or transformation for large-scale distribution. By minimizing waste and increasing the economic efficiency of food production, food processing has played a major role in raising our standard of living. Modern food technologies have given us easy access to a better, more varied, and sophisticated diet than was ever available to our parents, at least in our part of the world, as well as the possibility of exporting perishable foods to other countries.

Canning constitutes one of the greatest advances in industrial food processing. Its advantages for the preservation and transportation of foods are so well known that they need no elaboration. Sales of canned or bottled foods have soared over the years and in the United States alone they amount to more than $20 billion per year. The most important and delicate part of canning is sterilization, which is achieved either by heating canned foods at very high temperatures or canning foods in sterile conditions. Despite precautions and controls, a small fraction of canned goods does not reach the consumer at the required degree of sterility. This may result in food poisoning. Fortunately, most cases are mild, except when meats or other canned goods are contaminated by botulinum bacteria[a] whose toxin causes a very serious, sometimes fatal, disease (botulism). Another health hazard is associated with the deterioration of the materials (lead alloys) used in making the containers. The formation of voltaic couples can cause foods to be contaminated by toxic metallic ions.

Not entirely safe either are some of the more recent methods of food packaging. Large use is made of paper, plastics, and aluminum foil, and some of the substances used in these materials as stabilizers or antioxidants may be transferred to the food itself. Many plastic materials, for example, contain heavy metal derivatives as stabilizers, while polyvinyl chloride often contains its monomer vinyl chloride, whose carcinogenic properties have been clearly established. As the hidden dangers connected with the use of these materials become better understood through research by health agencies and by the producers themselves (as in the case of vinyl chloride), continued efforts are made to refine packaging laws and regulations and to improve packaging techniques.

Two additional techniques used for the preservation and transportation of food are freezing and freeze-drying. Neither involves chemistry to any marked degree, nor does yet another method of sterilization used by the food industry, namely, ionizing radiation. By eliminating pathogenic agents, this process enables foods to be preserved for a long time and thus to be transported and consumed a long way from the point of origin. As is always the case where radiation is concerned, this practice is a source of some worry to the health officials of the various countries that have adopted it. From extensive and careful tests, however, no evidence has emerged to suggest that this particular technique should be discouraged.

*Food Additives*

Many of the modern food technologies are based on chemical processes or chemical substances. Industrial food processing encompasses a variety of techniques, which range from injecting meats with proteolytic enzymes (meat tenderizers) to adding mineral salts, vitamins, and amino acids to ready-made cereal products, from oil refining processes requiring washings, hydrogenation, and steam distillation to the production of margarine, amounting to tens of millions of tons per year, and the production of starch from corn and potatoes, also reckoned in millions of tons. Food additives play an important role in the stabilization, preservation, and packaging of foods. As defined by Italian law (Health Ordinance of 31 March 1965), additives are "substances devoid of nutritional value or used for nonnutritional purposes that are added to the body or surface of foods at any stage of their preparation in order to preserve in time their chemical, chemicophysical, or physical properties, to prevent spoilage, or to impart to them, or enhance, particular characteristics of appearance, taste, smell, and consistency." Additives, in other words, are chemical substances deliberately introduced into foods for a very specific purpose. According to their function, they can be divided into various groups, but we will consider in some detail only two of them, *preservatives* and *coloring* and *flavoring agents*.

The group of preservatives includes *antimicrobial agents* and *antioxidants*. A number of *antimicrobials* are commonly used in food processing, each with a specific field of action. Benzoic acid and sodium benzoate are active at a $pH^b$ of 2.5–4 and are particularly effective against yeasts and bacteria, less so against molds. They are used to treat acidic beverages such as fruit juices. Derivatives of *p*-hydroxybenzoic acid act at a higher pH. Other antimicrobial agents are the unsaturated acids (e.g., sorbic acid) and salts of propionic and sulfurous acids. Salicylic acid and boric acid are falling into disuse owing to their marked degree of toxicity. The proposal of using antibiotics as food additives has not been favorably received. Food oxidation, whether biological or chemical in origin, results in the formation of various substances having an unpleasant taste, smell, and color and occasionally possessed of toxic properties to a lesser or greater degree. Oxidation may be caused by air, light, or microorganisms, with traces of heavy metals often providing the catalyst. Oxidation affects chiefly the fatty substances that are present in all foods. Thus such products as oil, butter, margarine, and fats in general are particularly susceptible to this type of deterioration. By inhibiting or slowing down chemical breakdown of fats, antioxidants play a very important role in food processing from both the technical and economic points of view. Most antioxidants are synthetic phenol derivatives, such as propyl gallate, butylated hydroxyanisole (BHA), and butylated hydroxytoluene (BHT).

Industrial processes of food preparation often result in color alteration. The original color is then restored by using the same coloring substances contained in the original products or synthetic dyes. The latter are more stable and offer a wider range of shades; furthermore, their supply is not affected by climactic or seasonal variations. For all these reasons the consumption of artificial colorings has soared in the last 30 years. It is estimated that in 1969 over 2,000 tons of synthetic dyes were used by the food industry in the United States alone. Food colorings are very similar in chemical structure to the dyes used in the textile industry, which include azide and sulfur derivatives of naphtol and other complex aromatic compounds.

Flavoring agents are the largest group of additives and perhaps the most profitable from a commercial point of view. There are at least 1,200 different flavors, two thirds of them synthetic. Over the years artificial flavors have come to replace natural flavors and spices for the same reasons we have mentioned with regard to food colorings— their supply does not depend on seasonal changes or the vagaries of agricultural production, and they are always available in the anticipated amounts and at fixed prices. The preparation of a flavoring agent requires a series of complex chemical processes. Just consider the fact that a natural aroma may contain from 100 to 200 ingredients, each in different amounts, and that the most prevalent substance is not necessarily the one that gives the greatest contribution to the quality of the aroma. Two of the most intriguing results in this field are the identification of the ingredient that contributes most to truffle aroma (formaldehyde dimethylthiacetale) and of more than 250 components of coffee flavor.

In addition to the additives mentioned, there are other compounds that are added to foods for a variety of reasons. One of them is *starch*, which is made from corn and potatoes. Millions of tons of starch are produced every year and used in countless food products such as jellies, jams, and ice creams. Then there are the *sweeteners,* or sugar substitutes, which are used in the preparation of foods for diabetics, dietetic foods, and so on. Best known are saccharin, dulcin, the cyclamates, and sorbitol. The first three sweeteners are under investigation by health agencies and their use has been restricted. *Thickening agents* are compounds that swell in liquids and are used to give consistency to certain foods. Fruit juices, ice creams, puddings, and jams owe their viscosity to natural thickeners like gum arabic, pectin, lecithin, and gelatin or to synthetic thickeners such as sodium carboxymethyl cellulose and polyphosphates.

### Advantages and Drawbacks of Additives

Quality is an all-important consideration in modern nutrition, and appeal is certainly a major quality factor. Our choice of foods is in fact determined to a large extent by some basic sensations such as flavor, aroma, and appearance (especially color). In turn, consumers' preferences determine the commercial value of a product. Hence the importance of those additives that preserve or enhance appealing characteristics and restore others lost or altered in the course of industrial food processing.

Some additives are indeed essential in modern nutrition, we are referring of course to preservatives (antioxidants and antimicrobials). By inhibiting or retarding deterioration, they allow food to be preserved for a long time, thus reducing waste due to spoilage and facilitating the transportation and distribution of food products. More controversial is the evaluation of all those substances that are added to foods just to make them more appealing, namely, sweeteners, thickeners, and coloring and flavoring agents. There is no absolute need for a food to be colored or for a pudding to be solid rather than syrupy. In the wake of the debates on the toxic effects of additives, numerous efforts have been made by the food industry to market processed foods that have not been restored to their original flavor or color, for example, grayish salami or colorless jams. Thus far, however, consumers' resistance to these innovations has been very strong. The use of this type of additives is therefore tied to a question of taste, hence to commercial considerations, rather than to real need. Clearly, an extensive educational campaign will be needed to break the vicious circle between the consumer's taste and the producer's economic interests.

There are reasons besides toxicity why additives may be objectionable. Additives may be used to mask defects of preparation and packaging, as has been known to happen. Worse yet, artificial colors can be used to revive aged goods and intense aromas can hide an incipient state of putrefaction. But for now let us consider the question of toxicity, meant here in its broadest sense. This is a very serious problem, the more so in view of the fact that food additives have become part and parcel of everybody's daily diet. Safety is a fundamental requirement for all such products, but in point of fact it is impossible to guarantee that a substance will be safe for everybody, in any amount, and under all possible conditions of use. Thus the only thing that can and must be done is to minimize the risk by careful testing and long-range studies. To this end, a special FAO-WHO committee has been gathering data for many years and encouraging research in the toxic effects of food additives. For each product under scrutiny three general parameters are measured: the *average lethal dose,* $LD_{50}$; the *maximum daily dose,* or the amount of additive that the organism can safely absorb every day; and *tolerance,* which denotes the maximum allowable amount in any food product.

Defining toxicity is not a simple thing, however. Animal tests are time consuming, costly, and not always applicable to man. Our intestinal flora can in some cases turn a harmless additive into a toxic substance. As intestinal microorganisms vary from person to person and an even greater difference exists between test animals and man, toxicity can manifestly become a subjective question. In addition, a number of people have special physiological conditions or organic diseases that render them particularly sensitive to certain chemicals. Last, many additives are not 100% pure and may contain harmful substances. All these considerations prompted the FAO-WHO committee to warn from the beginning that additives under study will have to undergo permanent surveillance in order to ascertain the possible occurrence of noxious effects and hence to modify their modes of use. What this means, in practice, is that all such products shall be considered on trial until such time as the pharmacological techniques at our disposal, now or in the future, will yield definite results. It is not surprising, therefore, that restrictions should be imposed on the use of some common additives and should also give rise to heated controversies that leave the consumer completely baffled and unable to make up his own mind on the issue.

Several products have come under suspicion. The FDA (Food and Drug Administration) has severely restricted the use of boric and salicylic acids and expressed reservations about butylated hydroxyanisole (BHA) and butylated hydroxytoluene (BHT), all of which are food preservatives. Hardest hit has been the group of coloring agents, which is scarcely surprising considering that synthetic dyes are generally cyclic compounds. Such molecules are very often mutagens or carcinogens. In addition, many food dyes have modifying side groups (e.g., amino or azido groups) that are suspected of enhancing carcinogenic effects. Some of the recently banned food colors include naphthol yellow S, Sudan IV, Erythrosine J, methylorange, and many other dyes whose toxic effects have lately received worldwide press coverage. Some flavoring agents that had been in commercial use for a long time were also found to be toxic and subsequently banned. Worth mentioning is safrole, a natural product long used as flavoring in some alcoholic beverages. It is the main component of sassafras oil, from which it is derived. Safrole was recently banned because accurate studies have revealed its marked carcinogenicity. (Frequently "synthetic" is taken as synonymous with "hazardous," "natural" with "healthful." But the example of safrole shows that this is not always true!)

With all such products, however, the difficulty lies in the evaluation of the dangers. It is not rare to find a difference of opinion even among people who are experts in the field and whose concern for public welfare is unquestionable. In the matter of additives the position of many health agencies is one of extreme caution. At the least hint of toxicity licenses are suspended or revoked. It is a questionable criterion, but before rejecting it out of hand we should consider its merits. Recall that in an analogous situation the United States was spared the thalidomide disaster only because an FDA researcher raised objections that seemed exaggerated at the time but later proved to be tragically accurate.

On the other hand, in an FAO publication, *Food and Nutrition,* we read that "American cancerphobia . . . is a disease as serious to society as cancer is to the individual."[13] With regard to cyclamate and food colorings the article says, "How can you feed a rodent amounts of cyclamate that would be equivalent to up to 1,300 bottles of diet soda a day and then generalize the results to man? Another example: If rodents are fed for many months a diet of which 3% consists of the food dye Red No. 2—as was done by the US Food and Drug Administration—how can the results be extrapolated to man, who receives less than 0.01% of the dye? It is, in fact, surprising that at 3% of the diet the white rats didn't turn red!" But there is no better example of the different criteria applied to the evaluation of food additives than what I would call "the saccharin mess."

## The Saccharin Mess

*The 18 April 1977 issue of* Chemical and Engineering News *had words of comfort for the readers who might be baffled by the saccharin issue: "If the Ringling Bros. Barnum & Bailey circus weren't in the middle of its annual springtime run in Washington, D.C., the current debate over saccharin might be seen as the Greatest Show on Earth."[14] This comment refers to the controversy sparked by the FDA's decision to ban the use of saccharin, one of the most common no-calorie sweeteners and practically the only one left after the ban of the cyclamates. The history of saccharin is not without its lighter moments. In 1907, to cut short a similar debate, Theodore Roosevelt officially declared, "Anybody who says saccharin is injurious is an idiot."*

*The current controversy revolves around a study conducted by Canadian scientists who fed saccharin as 5% of the diet [0.04 ounces per pound (2.5 grams per kilogram) of body weight] to two generations of laboratory rats. Among first-generation rats, 3 in 100 developed cancer of the bladder, while among second-generation rats, exposed to saccharin in utero as well as throughout their lives, the incidence of bladder cancer was 14 in 100. These findings prompted the FDA to apply to saccharin the so-called Delaney clause of the 1958 Food, Drug and Cosmetic Act, which states, "No additive shall be deemed to be safe if it is found to induce cancer when ingested by man or animal, or if it is found, after tests which are appropriate for the evaluation of the safety of food additives, to induce cancer in man or animal."*

*Reactions to the FDA's decision have been very strong, both in the United States and abroad, and have resulted in a controversy that in some respects epitomizes the great uncertainty surrounding the whole question of safety. At the governmental level, British health officials have voiced considerable skepticism over the Canadian results. They point to epidemiological studies conducted in England that show no correlation between the introduction of saccharin (which dates back to the beginning of the century) and cancer of the bladder. Japanese authorities, on their part, cite research carried out in 1975 at their National Institute of Health from which no evidence has emerged linking saccharin to cancer of the bladder in rats. Public reaction has been sharp and vocal. Diabetics and dieters have flooded government agencies, Congress, and health associations with letters and telegrams protesting the FDA's decision, and anybody with scientific qualifications has felt motivated to comment on it, not always appropriately.*

*One criticism leveled at the Canadian results concerns the purity of the saccharin used in the tests. "It is now known that the supposedly contaminant-free saccharin used in the Canadian experiments may not be pure after all. Absolutely pure saccharin . . . is not mutagenic in the Ames test. . . . The saccharin the Canadians used is. Thus, the possibility that cancer was caused by some yet unidentified impurity cannot be excluded as was originally thought."[15] The fact that the Ames test is negative for pure saccharin is, objectively, a strong point in favor of its supporters.*

*A second and more general criticism of the FDA's decision is based on the safety record of the substance. Critics point out that saccharin has been on the market since the beginning of the century and that in all this time, despite all kinds of tests, no evidence has emerged linking its use to any ailment whatever. On the whole, a valid argument. Toxicological tests—not the most rigorous, in*

*truth—were performed in Europe as far back as 1886. Some European work-*
*ers were given single 5-gram (approximately 0.2 ounce) doses of saccharin*
*without showing any adverse effects. In 1888 French diabetics were treated*
*with 5 grams of saccharin a day for 5 months, again with no ill effects. Later*
*on, President T. Roosevelt appointed a scientific committee to study the toxi-*
*cological properties of saccharin. The committee concluded that a daily dos-*
*age of 0.3 grams was absolutely safe, "and that levels above 1 gram per day*
*caused digestive disturbances." The first chronic toxicity study was performed*
*in 1951 on laboratory rats, which were fed up to 5% saccharin in their diets.*
*"This study did not reveal any reason to suspect saccharin was a carcinogen or*
*that it was harmful in any other way" (W. C. Lepkowski, "Saccharin ban goes*
*beyond issue of cancer,"* Chemical and Engineering News, *April 11,*
*1977). These conclusions were supported in 1955 by the Committee on Food*
*Protection of the National Academy of Sciences–National Research Council*
*(NAS-NRC), which confirmed saccharin's safety. In 1968 a report prepared*
*for the FDA by an ad hoc NAS-NRC committee set the maximum recommended*
*dose at 1 gram per day. This committee also recommended that long-term*
*studies on saccharin's cancer-causing potential be carried out, previous studies*
*having been judged inadequate and technically outdated. Toxicological tests*
*conducted on rats in 1972 showed that a level of saccharin equivalent to 1*
*gram per day for man presented no danger whatever, while at levels 100 times*
*higher there was a fairly high incidence of bladder tumors. In February 1977*
*the Canadian study began, with the results described.*

*Much of the current debate revolves around the Delaney clause and in par-*
*ticular around the term "appropriate" in relation to carcinogenicity tests.*
*Questions have been raised about the appropriateness of the tests on which the*
*FDA based its decision to ban the sweetener. Critics argue that the dose of*
*saccharin fed to rats by the Canadian scientists is about 1,000 times greater*
*than the amount normally found in dietetic foods and other artificially*
*sweetened products. To approximate the test dose, they point out, one would*
*have to drink 1,000 bottles of soda a day, or eat 210 pounds of toothpaste a*
*day, or drink 116 gallons of mouthwash a day. The American Diabetes Associ-*
*ation along with many other organizations, old and new, has come forth to*
*claim for diabetics the right "to drink diet sodas without fear."*

*Other critics add to the fire by comparing the Canadian saccharin study to*
*tests performed on substances that are normally found in man's diet. For exam-*
*ple, it can be shown that tryptophan, one of the amino acids essential to human*
*nutrition and found in all proteins, is a carcinogen when judged by the same*
*criteria used for saccharin. Following such criteria, "A glass of cow milk pre-*
*sents the same risk of carcinogenicity as a dietetic food containing 120 milli-*
*grams of saccharin." Thus "if the Delaney clause were to be applied to the*

*constituents of normal food, the FDA would have to ban meat, fish, eggs, milk, wheat, beans, and every other type of protein."*[16]

A last point is made by saccharin's supporters. The ban of artificial sweeteners will necessarily result in increased sugar consumption, which in turn means an increase in tooth decay, ischemia, gout, dermatitis, and obesity. It is pointed out that sucrose has been named by the Norwegian government as one of the substances whose use ought to be regulated.

The results of some recent toxicological studies generally support the view of saccharin's safety. Guinea pigs tested for 18 months at the University of Nebraska did not show any adverse effects. Studies on rhesus monkeys conducted over a period of 7 years also produced no evidence of toxicity.

Faced with such strong opposition, spearheaded by over 200 members of Congress, the FDA is considering a compromise solution, namely, banning the use of saccharin as a food additive and in cosmetic products (toothpaste and the like), but allowing it for medical purposes. This solution would have the advantage of eliminating the risk associated with the vast consumption of dietetic products and at the same time making saccharin available to people like diabetics, for example, who use it out of necessity.

There is a difficulty, however. To be classified as a drug, a product must be of some manifest benefit. And when it comes to defining saccharin's benefits, things become unclear again. The advantages deriving from the use of saccharin have been called "of the anecdotic type": There are people with a weight problem who show pictures of themselves before and after the cure; there are doctors who maintain that the use of saccharin is helpful in preventing heart disease but can offer nothing documentable. A thorough study by the Institute of Medicine of the NAS has not settled the matter. According to a spokesman, "The data on the efficacy of saccharin or its salts for the treatment of patients with obesity, dental caries, coronary heart disease, or even diabetes has not so far produced a clear picture to us of the usefulness of the drug." On the other hand, he adds, "There isn't any good evidence that saccharin causes human cancer either" ["Saccharin: A chemical in search of an identity," Science 196(4295): 1183 (1977)].

To complete the picture, mention should be made of a poll conducted by a prestigious scientific journal among highly qualified exponents of government health agencies and independent research organizations on the advisability of amending the Delaney clause. William Lijinski of the Frederick Cancer Research Center feels that any amendment to the clause aimed at making it less restrictive would result in very serious hazards to public health.[17] On the other hand, Frederick Coulston of Albany Medical College feels that the clause is "redundant, too rigid, and not needed by the FDA" and would like to see it repealed entirely. Since this is not politically feasible, the clause should "be *modified to fit updated toxicological concepts." In the specific case of saccharin and cyclamate, with the appropriate modifications to the clause "it would be possible to arrive at an acceptable daily intake for such chemicals," avoiding the total ban* (Chemical and Engineering News, *27 June 1977). Sidney N. Wolfe, of the Health Research Group, would also like to see the clause modified, but in a restrictive sense. In his opinion, no benefit, however great, can compensate for the risk, however small, of cancer.*

*Representative J. G. Martin, along with 200 colleagues, has proposed to Congress that an exception to the Delaney clause be made for saccharin. He, too, feels that the clause is anachronistic and not attuned to present conditions and ought to be modernized and changed. Aside from other considerations, it penalizes synthetic additives while ignoring the consumption of natural foods containing powerful natural carcinogens.*

*An unbelievable tangle. A veritable mess. I will leave saccharin to its uncertain fate without having been able, in the ocean of printed paper, to form an opinion of my own.*

### Additives in Animal Nutrition

Turning now to animal nutrition, we find that the use of synthetic products is far more substantial than in human nutrition. In Italy, for example, the annual consumption of additives and diet supplements amounts to about $40 million (at 1973 prices). Feed additives include amino acids (notably lysine and methionine), antibiotics, pigments, antioxidants, and coccidiostats. Such substances are added either for therapeutic purposes or to increase the nutritional content of feeds. Some antibiotics, for instance, promote weight gain in food animals. To quote some figures, in 1973 Italian stock farms consumed 400 tons of zinc bacitracin, 85 of chlortetracycline, 36 of oxytetracycline, and 60 of spiramycin—all of these are antibiotics, about 1,000 tons of coccidiostats[c] and 150 tons of nitrofurans. Urea is an important component of animal diet. It is estimated that in the United States more than 500,000 tons of urea are added every year, either directly or in solution, to silage and other fodder, which allows a saving of several million tons of soybean meal. We should recall that ruminants are able to use the bacteria in their stomachs to make proteins from simple nitrogen compounds such as urea.

The massive use of additives and diet supplements entails a degree of risk to human health. One danger, for example, is that the substances added to the feeds may not be eliminated before the animals are slaughtered. Hence the meat we eat may contain traces of nitrofurans or sulfonamides. Antibiotics present a special problem. The massive use of these drugs could replace susceptible microbes with resistant strains difficult to combat in the event of infection. This, in turn, would necessitate the use of different antibiotics, creating the promise of an endless race between microbes and drugs.

## Some Remarks on "Natural Foods"

So far I have discussed the products that chemistry has introduced in greater (fertilizers and insecticides) or lesser (colorings and so forth) measure into agricultural practices and food processing—all synthetic products and for the most part foreign to the "normal" cycles of soils and organisms. Considering their alarming diffusion and the host of detrimental effects I have described, it is scarcely surprising that they should be controversial, although opposition tends to be more vocal in those countries (generally the more affluent) whose people do not fully realize to what extent modern agriculture and nutrition rely on chemicals.

Health hazards, however, are not confined exclusively to synthetic products. Many "natural" foods (some say *all* foods) contain one or more substances that are potentially toxic. This type of danger is not only more insidious but much more difficult to confront in a rational way. Any discussion, even among competent people, can legitimately end in a stalemate. To give an example: A glass of whiskey or cognac may be considered by one expert as a pleasant drink, rather stimulating, and quite good at warming up a party. His more pessimistic colleague, on the other hand, may see it as a tenth of the lethal dose of alcohol. Alcohol intoxication, acute or chronic, is actually quite common. Furthermore, alcohol interacts synergistically with some drugs, a fact of which not too many people are aware. These are all good and sufficient reasons to consider the sale of alcoholic beverages as a serious danger to public health.

Let us come now to more down-to-earth foods. Everybody knows that prussic (hydrocyanic) acid is a poison, but a little-known fact is that bitter almonds, cassava, and sorghum may contain as much as 0.0008 ounces per pound (25 milligrams per kilogram); bamboo, as much as 0.0025 ounces per pound; and the various kinds of beans, 0.00008–0.001 ounces per pound. *Phaseolus lunatus,* a type of bean, caused a series of poisonings and killed seven people in Puerto Rico between 1917 and 1925. As for the oil of bitter almonds, it is so lethal that is has often been used for suicides and homicides. Clover contains saponins (natural substances named after the plant *Saponaria officinalis*) that can cause edema and respiratory arrest in cows. Lathyrism is a type of poisoning fairly widespread in India and Algeria, less so in Italy and France; it is due to akaloids of *Lathyrus sativus*, a common legume. Then there are allergies: Chocolate, coffee, tomatoes, almonds, tea, strawberries, beer, and spices are just a few of the many foods that can cause allergic reactions. These foods may have a *threshold effect*, that is, they may not cause a reaction until a certain quantity is eaten, which varies from person to person.

The list goes on. Fava beans are considered dangerous in certain areas of Sardinia. Shellfish often cause intoxication. Truffles can be fatal if eaten in large amounts. Some plants contain estrogens, and bananas contain powerful vasoconstrictors such as noradrenalin and serotonin (The latter is suspected of causing cancerlike heart ailments among some African tribes that eat a lot of bananas.) Several journals have reported that various cheeses contain physiologically active amines such as tyramine (Camembert, Parmesan, and Gorgonzola), putrescine (Camembert), and histamine (Camembert and others). Histamine, which can have some unpleasant effects, is found not only in cheese but in French wine and in smaller measure in Italian wine. Vitamin inhibitors are found in many foods—for example, antivitamin D in soybean meal, antivitamin E in beans and peas. Enzyme inhibitors have been discovered just about everywhere.

And that is not all. Some substances (e.g., selenium and some vitamins) are not only safe but essential below a certain level and are toxic above it. The situation is summarized in the preface of a book by E. M. Boyd: "The results [of studies conducted on a variety of foods] demonstrated that there are toxic and lethal doses of all pure foods."[18] This fact is the basic reason for so much controversy on food safety. Are bananas dangerous? Normally not, unless one eats too many for too long. The same for carrots. But how many is "too many"? How can we set a limit when we do not know what else people eat beside bananas or carrots, when we do not understand the interactions between different foods, and when only some foods have been studied with modern techniques, while we know nothing scientifically valid of others? A varied diet is clearly preferable in this respect. It is not by chance that the toxic effects of certain foods are first observed on farms where the animals are fed the same diet for long periods of time, a practice that enables harmful substances to accumulate.

In this whole question of safety, therefore, it is very difficult to give the definite answers that the layman has come to expect from science. Since so much here is a matter of opinion, most experts can only make general recommendations and propose provisional criteria, aware as they are that everything has a degree of toxicity and that what counts is the relative scale of priority that must be established between one toxic substance and another, between one pollutant and another, between overall drawbacks and advantages. In this framework there are more than enough grounds for controversy and for the uncertainty that often surrounds the experts' evaluation of a food product. This conclusion, I fear, is bound to disappoint all those people who always ask science for answers in black and white, rather than in the more realistic shades of grey.

# 3

## *Chemistry and Health*

No field better lends itself to controversy in the matter of chemistry's benefits and risks than medicine. Every medicinal substance can be praised or damned. Morphine, so useful in hospitals and in surgery, is one of the calamities of modern time. Penicillin saves millions of lives, but in allergic patients may cause anaphylactic shock and even death. Aspirin, the pain reliever so extensively and indiscriminately used for many years, can cause serious troubles to the digestive tract. And of course there are the tragic episodes like the thalidomide disaster, where the benefits expected from a drug were far exceeded by its dangers.

It cannot be denied, however, that some diseases have been conquered and that the average life span has substantially increased. Life expectancy for a child born in 1900 was 47 years. It rose to 56 in 1920 and to 70 years in 1970. (These figures apply only to the more technologically advanced nations.) Such a drastic change does not depend, of course, on any single factor, but on a combination of things: a general improvement in the standard of living, better and more abundant nutrition, postnatal care, decrease in infant mortality, and medical advances.

One of medicine's greatest contribution to man's health is the development of vaccines. Through immunization, some of the world's most dreaded diseases have become preventable. The most recent accomplishment in this field is the worldwide eradication of smallpox, which was officially announced by the WHO a few years ago following the elimination of the last reservoirs of the disease in Ethiopia. Other great medical advances have been the discoveries of vitamins, sulfa drugs, and antibiotics. If success is measured in terms of decreased mortality rate for infectious diseases and increased life expectancy, the medical sciences can certainly claim to have made great strides in the more technologically advanced countries. In 1900 about 500

Americans in 100,000 died of infectious diseases; at present, the number is as few as 50. Syphilis and tuberculosis have been vanquished, diphtheria is no longer dreaded, and gastrointestinal infections are much less common and largely curable.

All these accomplishments have inspired technological man with a great deal of faith in medical science. There are many, however, who not only question medicine's contribution to the overall decline in infectious diseases, but view the ever increasing and unrestrained consumption of drugs and the development of a public health industry as very real dangers for modern society. Here is what Ivan Illich, previously mentioned with regard to the DDT controversy, writes in his essay attacking modern medicine:

The infections that prevailed at the outset of the industrial age illustrate how medicine came by its reputation. Tuberculosis, for instance, reached a peak over two generations. In New York in 1812, the death rate was estimated to be higher than 700 per 10,000; by 1882, when Koch first isolated and cultured the bacillus,[a] it had already declined to 370 per 10,000. The rate was down to 180 when the first sanatorium was opened in 1910, even though "consumption" still held second place in the mortality tables. After World War II, but before antibiotics became routine, it had slipped into eleventh place with a rate of 48. . . . The combined death rate from scarlet fever, diphtheria, whooping cough, and measles among children up to fifteen shows that nearly 90 per cent of the total decline in mortality between 1860 and 1965 had occurred before the introduction of antibiotics and widespread immunization.[1]

There is no question that health care is big business. Consider the United States, for example. The American pharmaceutical industry employs 140,000 people, distributed in 1,300 industrial units. In 1970 Americans spent $67 billion for health care, about $6 billion for drugs alone. In 1974 this figure was already in excess of $9 billions. Turning to Europe, in 1974 the German Federal Republic produced $3.8 billion of pharmaceutical products, France $2.3 billion, and Italy $2 billion (OECD figures). According to Farmunione statistics, pharmaceuticals constitute about 1.2% of Italy's GNP (gross national product).[b] In 1976 total health expenditure in Italy added up to $9,800 million, or 6.64% of the GNP. Per capita health costs amounted to $150, and per capita drug costs to $36; total expenditure for drugs (at retail prices) was $2,400 million (15.6% of the total health expenditure). For Italy's export-import trade balance see table 3.1. In 1974 the Italian pharmaceutical industry (including production of preparations for human and veterinary use, galenicals, and

**Table 3.1**
Italy's import-export of pharmaceuticals

| Branch | Business volume ($\times$ \$10$^6$) | % | Imports ($\times$ \$10$^6$) | Exports ($\times$ \$10$^6$) |
|---|---|---|---|---|
| Pharmaceutical industry | 25 | 100 | 4.5 | 5.5 |
| Active principles | 5 | 20 | 1.25 | 3.8 |
| Specialties for human use | 20 | 80 | 3.5 | 1.8 |

Source: Report of the Division of Industrial Chemistry of the Italian Chemical Society presented at the Congress of Merano (June 1978).

**Table 3.2**
The European pharmaceutical market

| Country | Sales ($\times$ \$10$^6$) | Items sold ($\times$ 10$^6$) | Average price (\$) |
|---|---|---|---|
| Italy | 13 | 1,627 | 0.8 |
| France | 20 | 1,828 | 1.25 |
| England | 9 | 460 | 1.65 |
| German Federal Republic | 30 | 1,256 | 2.35 |

Source: *Panorama Farmaceutico* 1(1) (1978).

basic chemical compounds) employed an estimated 62,500 people, 54,100 in the Center-North and 8,400 in the South. In the same year fixed investments were evaluated at \$90 million (\$75 million in the Center-North, \$15 million in the South).

It may be interesting to examine in some detail the quantitative and qualitative aspects of drug consumption in a few European countries. Table 3.2 gives some figures concerning the European pharmaceutical market. Italian products cost an average of \$0.8 versus \$1.25 for French products, \$1.65 for English products, and \$2.35 for West German products. The English are modest consumers of drugs, relatively speaking, while the French buy slightly more medicines than the Italians. Table 3.3 which illustrates drug consumption in Europe according to pharmacological groups, reveals some intriguing national differences. Allowing for the difference in population, it appears that the Italians are the strongest consumers of liver and ulcer preparations and the Germans of diabetic products and sedatives. The French seem to be chiefly concerned with arteriosclerosis and the Spanish buy mostly antibiotics. These figures, which show a remarkable difference in behavior among Europeans, are clear evidence of a certain lack of health education.

## The Development of Pharmacology
According to *Chemistry and Economy*,[2] an accurate study of the American Chemical Society, the key steps in the growth of the pharmaceuti-

**Table 3.3**
Consumption of pharmaceuticals in Europe (in percent)

| Pharmacological group | Italy | Belgium | Holland | France | GFR | United Kingdom | Spain | Sweden |
|---|---|---|---|---|---|---|---|---|
| Ulcer preparations | 42.5 | 0.7 | 1.4 | 11.6 | 24.3 | 6.5 | 13 | – |
| Liver preparations | 49.5 | 0.9 | – | 18.2 | 21.7 | – | 9.3 | 0.2 |
| Diabetes preparations | 11.7 | 2.9 | 4.5 | 16.0 | 51.3 | 4.0 | 5.6 | 3.6 |
| Arteriosclerosis drugs | 25 | 1.2 | 1 | 30.5 | 21.9 | 3.4 | 14.2 | 2.5 |
| Coronary dilators | 10.5 | 6.1 | 1.3 | 35 | 39.2 | 0.9 | 5.7 | 0.9 |
| Antibiotics | 22.6 | 7.8 | 4 | 20.4 | 9.4 | 7.2 | 26.1 | 2.2 |
| Sulfonamides | 6 | 7 | 5 | 9 | 43.2 | 21.6 | 5 | 3 |
| Tranquilizers | 7.2 | 7.9 | 5.2 | 27 | 36.3 | 6.6 | 6.9 | 3.6 |
| Hypnotics and sedatives | 10.4 | 4.2 | 6.1 | 21.9 | 39.1 | 10.8 | 3.2 | 3.7 |
| Antidepressants | 5.9 | 6.4 | 5.4 | 25.4 | 20.7 | 6.4 | 2.1 | 4.1 |
| Total market | 17 | 4.3 | 3.4 | 24.3 | 28.2 | 6.5 | 12.4 | 3.6 |

Source: *Panorama Farmaceutico* 1(1) (1978).

cal industry are associated with sulfa drugs, antibiotics (penicillin, streptomycin, tetracyclines), isoniazid for the treatment of tuberculosis, adrenocortical hormones for allergies and rheumatism, *Rauwolfia*[c] and Veratrum[d] alkaloids for the treatment of hypertension, diuretics based on organic mercurials and thiazide, chlorpromazine and other derivatives for the treatment of anxiety, antihistamines, insulin and other products for the control of diabetes, oral contraceptives, and anesthetics.

Up to 1920 most drugs were natural in origin. Malaria was treated with quinine, anesthetics consisted of extracts of coca or opium, and infusions of all types were the main tools at the disposal of patients and doctors alike. Iatrogenic diseases[3] (that is, diseases caused by improper use of drugs) had not yet assumed the epidemic proportions they would reach in our time, but it is also fair to say that diseases currently considered trivial, or at least nonfatal, such as typhus, influenza, pneumonia, and diphteria, were then as feared "as the plague."

In the 1920s synthetic products began to be used in increasing numbers, and the early 1930s saw the advent of sulfa drugs, which marked a turning point in the treatment of infectious diseases.[e] In the case of the much dreaded pneumonia, to give just one example, the mortality rate before the introduction of sulfa drugs had been 2–4 patients in 10; afterward it fell to 1 in 15.

The discovery of antibiotics was another great advance in chemotherapy. The first of these drugs, and still one of the most widely used, was penicillin, discovered by A. Fleming.[f] What Fleming observed was that certain molds inhibit the growth of bacteria. Responsible for this effect is a particular substance secreted by such molds, namely, penicillin. Fleming's observation provided the impetus for a tremendous amount of research in the field. By extending the original criterion to other molds, investigators discovered many other antibiotics that have proved effective against a variety of microorganisms. As a result of this work, most diseases caused by bacteria can now be treated by chemical means. Tuberculosis, for one, is no longer the scourge it used to be. Although it may already have been on the decline before 1900, as Illich maintains, the fact remains that while in 1900 the tuberculosis death rate in the United States was 200 in 100,000 people, in 1954 it had fallen to 10 in 100,000 and in 1963 to 3 (American Chemical Society data).

In those countries where infectious diseases have been brought under control and the average life span has increased, a greater percentage of people have begun to suffer from ailments that were not so prevalent in the past. This change is reflected in the enormous increase in cardiovascular, cancer, and cerebrovascular deaths. Table 1.4 illustrates how the leading causes of death have changed from 1900 to 1960 in a highly industrialized country like the United States and the preeminent role assumed by cardiovascular disease. A number of drugs have been developed for the treatment of heart disease and of high blood pressure in particular, including various derivatives of reserpine, a natural alkaloid, and the synthetic compound guanethidine. A third remedy for hypertension more recently introduced into therapy is $\alpha$-methyldopa. And since doctors are often faced with the opposite problem, that is, a fall in blood pressure brought on by hemorrhage or shock, numerous vasoconstrictive drugs derived from the angiotensins have been developed to treat this condition also. Hypertension and certain kidney and liver disorders can cause retention of water and salts, an ailment that in its most serious form is called edema. Diuretic drugs are quite helpful in some cases. In the early days of the medical science diuretics consisted of natural products such as theophylline and digitalis,[g] but they were later superseded by organic mercurials and thiazide derivatives among others. The treatment of diabetes has been revolutionized by insulin, insulin zinc suspension, and other products.

The last two decades have seen a tremendous increase in the use of drugs that affect the central nervous system. This is an extremely difficult field for therapy because the accurate diagnosis of an ailment, the choice of medications, and the evaluation of their effectiveness depend to a very large extent on the subjective impressions that the physician receives from the patient's equally subjective description of his feelings. When we move from serious and clearly definable diseases like schizophrenic or manic depressive syndromes into the realm of neuroses or neurasthenia, the use of a drug frequently elicits a very complex response in which suggestion is a significant component. The current use of psychotropic drugs[h] (sedatives, stimulants, tranquilizers, hypnotics) is enormous and not always justified. All too often such drugs as reserpine, benzodiazepine, chlorpromazine, and phenothiazine derivatives are the best-selling items of the huge pharmaceutical companies that produce them.

Considerable progress has also been made in the field of analgesics. The problem here was to replace powerful natural pain killers like morphine and its derivatives with substances that would be just as effective in alleviating the terrible suffering that often accompanies surgical intervention without the equally terrible side effect of habituation. Narcotic addiction is a social plague of major proportions, and

its spread has been favored in part by the fact that morphine derivatives play a legitimate therapeutic role for which adequate alternatives
do not as yet exist. Although severe restrictions have been imposed on
their use, they are never tight enough to prevent addicts from obtaining the drugs from complacent doctors, by forging prescriptions,
and so forth. The search for nonaddictive pain killers, which has resulted in the introduction of pentazocine and cyclazocine among
others, is related to the study of products that can be of help in the
treatment of addiction, that is, in the rehabilitation of those addicts
who either voluntarily or forcibly are weaned from their dependence
on narcotics. Some products, notably methadone, have proved
beneficial in alleviating the withdrawl syndrome.[4] For the moment,
however, there is a great deal of uncertainty in this field, owing to the
significant risk connected with the use of certain medications and antidotes. (Methadone is a case in point. All too often methadone addiction replaces, or is added to, morphine or heroin addiction.)

Research in the field of antihistamines[i] has had very positive results,
first with the clarification of histamine's role in the complex series of
phenomena that cause allergic disorders, and then with the preparation of drugs that afford, if not a cure, at least some relief from the
most unpleasant symptoms. The pheniramine group has proved one
of the most effective.

Steroid compounds are a group of drugs that have proved quite
effective therapeutically when used as antiinflammatory agents, sexual and adrenocortical hormones, anabolic agents, and contraceptives. They are useful in female and male menopause, the treatment
of breast cancer, and disorders of the female hormonal cycle. Cor-
tisone drugs are the most widely used antiinflammatory agents, but
their adverse effects, which are becoming increasingly more apparent, have prompted research and production of antiinflammatory
agents of a nonsteroid nature. The use of steroid hormones as oral
contraceptives deserves special mention because of the importance
they have assumed in our society. Although controversial from many
points of view, they are currently being used by millions of women all
over the world.

*Drawbacks of Drugs*

The dangers connected with the use of drugs form a very complex
problem. There are a number of reasons why an accurate assessment
of these dangers is so difficult. To begin with, while it is relatively easy
to test the effects of a chemical agent on animals, the same cannot be

said for man. Except in a few well-understood cases, cause-and-effect relations are very difficult to establish. The effect itself is often unclear, let alone the causes. Furthermore the response to a drug is often confused by subjective impressions and contradictory feelings. In the second place, drugs are a lucrative business in the technologically advanced countries. Advertisement there is massive, it is estimated that in 1972 the American pharmaceutical industry spent for commercial promotion in excess of $4,000 for each of the more than 350,000 doctors in the country. The benefits of a drug are often overstated by its producers, and dissenting voices have a hard time making themselves heard in the chorus of praises. Finally, in the case of some drugs it is not only commercial or medical factors that give rise to controversies but ideologies as well. A case in point is the contraceptives, whose evaluation is often overlaid with ethical or religous considerations. Given the complexity of the problem, I will concentrate on a few cases that are not only important in themselves but representative of the general situation.

## Iatrogenesis

The most important of the ills brought by developments in medicine is *iatrogenesis,* or the pathogenesis due to excessive consumption of drugs.[3] Much has been said and written about this problem. One of the best known essays on this issue, and certainly the harshest, is the work by Illich already mentioned.[1] Basing his case on a voluminous amount of data, Illich stigmatizes the widespread drug abuse occurring in industrialized countries as a veritable social disease. People take drugs that, although presumed to be harmless, are in fact unsafe, have exceeded their expiration date, or are not needed. The problem is compounded by the fact that many of these preparations, when taken at the same time, involve a considerable risk of synergistic effects. Many drugs (or, more generally, many substances) act one way when taken alone and another way when the organism is concurrently being subjected to another substance. In other words, a harmless substance may become toxic when mixed with another. And since there are countless possible combinations, the experimental approach to the detection of these effects is very difficult, if not impossible. Alcohol and cigarette smoke, for example, are known to produce synergistic effects (see chapter 8). Mixing alcohol with tranquilizers can also have very serious, and sometimes fatal, consequences. In the United States 3–5% of hospitalizations are due to a bad reaction to a medication; 18–30% of hospitalized patients have pathogenic drug reactions that result in the worsening of their clinical conditions and lengthier hospital stays. Useless surgery—again according to Illich—also falls in the category of iatrogenic diseases. He holds that many surgical interventions are unjustified, being due to diagnostic errors or other motives.

Illich presents some convincing arguments in support of his thesis of a widespread drug abuse. Drugs affecting the central nervous system are the fastest-growing group in the US pharmaceutical market and constitute 31% of total sales. Since 1962 the consumption of substances likely to cause habituation or addiction (notably tranquilizers) has increased 290%, whereas in the same period alcohol consumption has increased 23% and narcotics use 50%. Apparently this phenomenon has more to do with the medical profession than with the political system. The first major trade agreement between China and the western pharmaceutical industry (1974) involved chiefly tranquilizers.

Iatrogenesis is undoubtedly a very serious drawback in the development of medicine and pharmacology. As a social phenomenon, however, it goes beyond the confines of our discussion, which is meant to be a risks-versus-benefits evaluation of individual products.

## The Problem with Antibiotics

An essential requirement for an antibiotic is an extremely selective toxicity, or the ability to be toxic to the pathogenic organism but not to the host. Thus, its toxic effect must be directed toward a specific biochemical characteristic of microorganisms rather than toward those essential life processes that are common to microorganisms and higher organisms. Understandably, the margin of safety is rather small. There are some antibiotics, however, that exhibit this selective toxicity by being toxic to injurious organisms but not to humans (relatively speaking—no substance is completely safe and much depends on dosage). Penicillin, for example, acts mainly by inhibiting the synthesis of new cell walls in bacteria or molds. Single microorganisms have cell walls that are peculiarly their own and differ in important ways from those of higher organisms. Hence the side effects on the host are mild, except of course in the event of allergic reactions, which can be very serious; some people become so sensitive to penicillin that they may undergo anaphylactic shock.

The risk to the host is much higher when the antibiotic affects more fundamental processes in the chemistry of living organisms. Such is the case with chloramphenicol, tetracyclines, and streptomycins, all of which interfere with crucial steps in protein synthesis. Since the biochemical processes involved in protein synthesis are basically the same in bacteria and humans, these drugs may interfere with the activity of both the bacterium and the host. Chloramphenicol, for example, inhibits the synthesis of antibodies and may also disrupt the production of red cells by the bone marrow. Other antibiotics, such as the polymyxins, may damage human cells by altering their lipid content, and still others attack nucleic acids in both bacteria and their hosts.

In addition to these side effects, which are closely tied to particular antibiotics, there are others of a more general character. The digestive tract contains an extremely important microbial flora that regulates various metabolic processes connected with the absorption of food and expulsion of waste. Besides damaging the patient because of its intrinsic toxicity, an antibiotic may disrupt the intestinal flora, thereby causing stomach disorders, as the result of protracted use. It usually takes quite a while to restore the normal biological activity of the digestive tract.

Another serious consequence of the excessive use of antibiotics is the development of resistant strains of microorganisms. It has been found that after a time the therapeutic effectiveness of all antibiotics tends to decrease. To give an example, while in the 1950s penicillin was effective against most of the bacteria that could be isolated from the environment, now the opposite is true: A great many of the bacterial strains that are isolated prove to be resistant to penicillin. Most of the gonorrhea cases among American Viet Nam veterans did not respond to penicillin G, a drug that used to be effective against this type of infection.

There are roughly three ways by which bacteria can become resistant to a particular antibiotic. A first defense mechanism is based on the ability of some microorganisms to produce enzymes capable of destroying the antibiotic chemically. In the case of penicillin, for example, some bacterial strains have succeeded in synthesizing a special enzyme, called penicillinase, that breaks down the penicillin molecule into inactive components. A second mechanism for the development of resistant strains is based on selection. In every microbial colony, which consists of millions of microorganisms, there are a few members whose biochemical makeup enables them to resist the action of a particular antibiotic. While the susceptible members die, the resistant strains survive and multiply until they become the majority in the colony. In the long run such a selection mechanism renders the antibiotic practically ineffective against that particular type of microorganism. There is yet a third and more insidious mechanism by which bacteria become immune to antibiotics. It has been shown that resistant strains are able to transmit their immunity genetically to members of the same species, as well as to members of different species. Since this process occurs by the actual transfer of genes (on plasmid DNA), such a trait can be handed down to successive generations.

As a result of these processes, once powerful antibiotics frequently prove ineffective and must be replaced with other antibiotics or mixtures of them. Because these drugs generally have some toxic effects and are administered to organisms already weakened by disease, the risks they pose cannot be overstated. This problem is most serious in hospitals since bacteria that have withstood thorough sterilization procedures as well as antibiotics are particularly hard to control. Infections by such virulent strains tend to be quite intractable and are a major factor in many deaths.

The emergence of resistant bacterial strains is undoubtedly a most alarming development and the main reason behind the endless search for new antibiotics. We just cannot afford to be unprepared in the event of epidemics caused by virulent bacteria that through selection or other mechanisms have become immune to all known antibiotics. It should be noted, moreover, that this is not a case of personal choice. It is not enough *personally* to give up the use of antibiotics except in cases of absolute necessity. The type of infection we may contract does not depend on the history of the host, but on the history of the infecting microbial strain, which is determined by the antibiotics added to animal feeds and used in therapy, for sterilization procedures, and so forth.

## Contraceptives

*No drug has done more to change man's habits and to create the conditions for even more radical changes than "the Pill." Oral contraceptives were developed between 1955 and 1960 and their story has been a complicated one. I should start by saying that birth-control research could hardly have been carried out in countries with a strong Catholic influence. With regard to Italy, a 1927 health regulation—article 163, only repealed in 1975 by an act of the Constitutional Court—stated that "approval shall be denied: ...(3) when the product has, or may be presumed to have, contraceptive properties or otherwise meant to disrupt the physiological course of gestation, or when its use may be in any way offensive to morality and accepted standards of behavior." The connection between contraception and public morality appeared consistent to the 1927 legislator and continued to be upheld in the 30-year period from 1945 to 1975. The fact that the term "contraceptive" could not be written on such preparations forced Italian manufacturers to disguise them under the strangest wordings, thereby effectively limiting their use. Up to 1975, for example, a contraceptive drug may have borne the following prescription: "Indicated for menorrhagias of puberty and menopause, primary and secondary amenorrhea, sterility, deficient genital development, early pregnancy diagnosis, temporary block of gonadotropic activity and ovulation, regulation of the menstrual cycle." Obviously it was not easy to see through the disguise and make the right deductions. Furthermore, such a prescription was wholly inadequate to explain the use of a product that requires great precision, no lapses, and a certain level of education. Understandably, this situation did not encourage academic and industrial researchers to carry out studies on contraceptives. In a social climate in which the promoters of birth control devices were threatened with arrest it was hardly possible to apply for government grants for contraceptive research, and quite unlikely that pharmaceutical companies would invest much capital in the development of products that—by law—could not be marketed.*

*Nor was this situation peculiar to Italy alone. Similar regulations were also in effect in France and, generally speaking, the issue of birth control was frowned upon or at least ignored in all European countries, Protestant and Catholic alike. Even in the United States, as C. Djerassi wrote in 1970, "Fifteen years ago this was virtually a taboo subject."[5] He also points out that "from the late 1950s until the early 1960s the US Government spent very little on the development of new birth control agents. The overwhelming portion of the cost of developing the oral contraceptives was met by three pharmaceutical companies."*

*It is worth recalling how oral contraceptives became available to mankind. The search for something that would curb female fertility is as old as the world. But the first scientific approach to the problem dates from the 1950s, when a thorough understanding of the role of hormones in the physiology of ovulation and reproduction led to the development of hormonal contraceptives. The basic principle of hormonal contraception is to simulate, by the administration of hormones, the physiological condition that prevents a new conception during pregnancy.*

*We owe the Pill to three scientists linked by events in a story that is a mixture of chance, intuition, and luck. The story begins with a courageous and eccentric chemist, Russel Marker, who heedless of his colleagues' advice walked the length of Mexico extracting active agents from plants in the hope of finding a convenient source of progesterone, a female hormone that mediates all processes connected with conception and pregnancy. At that time (around 1937) progesterone was one of the rarest and most expensive substances in the world. About one thirtieth of an ounce (1 gram) of it could cost as much as $200, but obtaining even that amount was not a simple matter. Progesterone was successfully used to prevent habitual miscarriage; the only limitation to its use was its rarity. Marker and his helpers examined over 400 plant specimens in a* *search that for years was fruitless and frustrating. At last, in a remote region inhabited by native tribes, they came upon a Mexican agave,* cabeza de negro, *whose roots were found to contain diosgenin, a substance from which progesterone could be easily produced.*

*Marker was not able to convince anyone of the importance of his finding. After applying unsuccessfully to Parke Davis for financial help, he gave up his professorship at Pennsylvania State College and moved to Mexico City. There, helped solely by Mexican peasants and with rudimentary equipment, he succeeded in synthesizing an amount of progesterone that for the time was very impressive. With about a pound of the precious substance wrapped in a newspaper he presented himself at the acquisition office of a Mexico City firm, Hormone Laboratories. Thanks to the contribution of its new partner, that small company became Mexican Syntex, a large industry that has had a monopoly on progesterone production until the present time. Then Marker disappeared without leaving his address in search of new adventures. Marker, incidentally, had never written down an accurate description of the chemical process he had used to turn diosgenin into progesterone, and his successors at Syntex had a devil of a time reconstructing the procedure from the vague accounts of his native coworkers. Syntex currently produces most of the raw material for steroid hormones (by a process that is still essentially Marker's own).*

*The second actor in our story is Gregory Pincus, a biologist whose main field of research were the relations between hormones and cancer and between mental illness and sexual anomalies. While Marker's discovery was the fruit of an obstinate courage, in addition to essential scientific intuition, Pincus's role was determined in large part by chance. On a winter evening in 1950 he was invited to the house of a physiologist who was deeply involved in the problems of birth control. Margaret Sanger, founder of the American birth control movement, was also present. The question of contraception came under discussion and the general consensus was that current methods were largely inadequate to cope with the problem. The latest mechanical device, the diaphragm, had proved unsatisfactory and scientists seemed unable to come up with a reliable and convenient way to prevent conception.*

*At the end of the evening Pincus accepted $2,100 to initiate contraceptive research. As told in the "International report on birth control" published by* Life *in 1967, during the solitary 180-mile car ride back to his home Pincus thought about the problem and had the basic idea that would eventually lead to the birth-control pill. During pregnancy a second conception cannot occur because ovulation, or the release of a mature egg by the ovaries, is inhibited by the presence in the female body of large amounts of progesterone, a hormone that is secreted in abundance during gestation.*

*Through Marker's efforts progesterone was readily available. It now remained for Pincus to test its effect on nonpregnant females. Together with a collaborator, he proceeded to administer progesterone pills to female rabbits and rats. Despite free and lively sexual activity, no pregnancies occurred. These encouraging results in laboratory animals opened the way to experimentation on humans—not a simple matter at all under the circumstances. To begin with, women are unlike female animals both psychologically and physically: They are always sexually receptive, and they have a particular cycle. Thus laboratory tests are not necessarily applicable to women. In the second place, asking women to volunteer for the experiment was contrary to medical ethics. Although it is not unusual to have new drugs tested on prisoners who have nothing to lose and offer themselves willingly, in this case it was clearly out of the question. Moreover, should the experiment prove unsuccessful, serious consideration had to be given to the possibility of inducing abortions; pregnancies initiated by the failure of the pill could not be allowed to continue, because the abnormal physiological conditions created by the administration of hormones might affect fetal health and physical integrity. A very difficult problem indeed, as is often the case when there is a need for human experimentation.*

*Enters now the third actor in our story, J. Rock, a professor of gynecology at Harvard University who was also experimenting with progesterone at the time but for different reasons; his objective was to find a treatment for female steril- ity. Rock's problem was to understand why some women, apparently normal from the hormonal point of view and capable of normal ovulations, could not conceive. The only detectable anomaly was the fairly small size of the fallopian tubes. Having observed that both the uterus and the fallopian tubes become enlarged during pregnancy, Rock thought of inducing pseudopregnancy by the administration of progesterone and estrogens. Some of his patients did in fact conceive and carry out normal pregnancies after this treatment, the only draw- back to which was the occurrence of nausea, headaches, and the various other symptoms of pregnancy.*

*Pincus and Rock agreed to try a new experiment. Instead of a mixture of progesterone and estrogens, Rock would give his patients progesterone only and the treatment was to last 21 days starting from the fifth day after menstruation. The results were quite good. Rock's patients felt much better than they had felt with the mixture of hormones, did not develop pregnancy symptoms, and their menstrual periods remained relatively normal. Furthermore, menstruations occurred regularly and tests showed that in most cases ovulation had been sup- pressed. Some of the patients subsequently conceived, which was Rock's objec- tive. As for Pincus, he saw a confirmation of his theory. Progesterone did act as an antiovulatory agent in women as well as in animals. Unfortunately, it had proved effective only in 80–90% of the cases, which was not good enough for an anovulant to be used for birth control. This partial failure was due to the fact that progesterone is more easily absorbed when administered by injection than when taken orally. The choice now was between inhibiting pregnancy by means of progesterone injections and looking for a progestin that would act orally. Pincus chose the latter alternative. Out of about 200 products tested on rabbits and rats a few were found that met all the requirements. They were effective orally, in small doses, and appeared to be completely reliable, at least in animals. The first type of pill, called Enovid, contained a progestin, norethynodrel, and a very small amount of the estrogenic hormone mestranol.*

*The next step was to test the drug on humans. Let us summarize briefly what was known at the start of the experiment: Some progestinic compounds had shown 100% effectiveness in animal tests, whereas progesterone was only 85% effective in inhibiting ovulation in women. The newly synthesized progestins, including Enovid, had never been tested on humans, and, moreover, their long-range effects on animals were not known. In other words, nobody knew whether adverse effects would show up in subsequent generations.*

*Suitable volunteers were found in Puerto Rico among the residents of a low-income housing project on the outskirts of San Juan. The women chosen for the test were all under 40, with more than 2 children and determined not to have any more. Some details taken from the "International report on birth control" will serve to illustrate the social milieu from which Pincus selected the first women to try the Pill. One of the volunteers was a woman 30 years old who had already had 10 children and whose husband was an alcoholic. Another had 5 children and a husband affected by serious psychological disorders.*

*The women's lack of education did prove a problem during the experiment. One of the difficulties, for example, was to impress upon them the need to take the Pill every day without fail. Despite the inevitable mistakes, the results were extremely satisfactory. The new birth-control method proved to be almost 100% reliable.*

*Following the Puerto Rican tests, the Pill spread throughout the United States, then to Europe and the rest of the world. The extension of its use to all women (recall that the original tests involved solely women below 40 and with at least two children) came about through the initiative of individual physicians. In a very short time the Pill became available to women who had never borne children and to the very young. The change in social mores that has occurred in the last few years has completely outpaced all pharmacological studies on the Pill's long-range effects, and today it is prescribed to girls who only 10 years ago would have been called children. Another reason for concern is the question of dosage. At the beginning the amount of hormones contained in the Pill was calculated to ensure complete effectiveness. Such a dosage amounted to an* overdosage *in individual cases. Only later on was this problem avoided (by decreasing the dosage, producing pills of different dosages, and introducing the so-called minipills).*

*Not enough time has elapsed for us to know about the possible genetic consequences of the Pill. (The Pill was developed in 1955 and approved for use in*

*1960.) We can only extrapolate to humans from the results of animal tests, but, as I said, women have a unique reproductive physiology. In the case of the Pill, however, absolute safety and the total lack of side effects were not the primary consideration. The benefits of the Pill were expected to far outweigh the more or less theoretical risks. At present millions of women take the Pill and hundreds of thousands of doctors prescribe it. Furthermore, many women take hormonal preparations on the advice of girlfriends or boyfriends without consulting doctors. Such widespread and indiscriminate use of the Pill raises some questions. What will its effect be on a woman who has not yet conceived? Or on a girl who has just started ovulating? More generally, how will the complex female physiology be affected by the protracted use of hormones and the concurrent suppression of ovulation? And what about future generations? In the case of the Pill it is clearly a question of accepting the risks for the sake of its considerable benefits. It should be noted, moreover, that the fact that its use raises questions does not necessarily mean that the danger is real, or that the health hazards are serious, or that its introduction was a mistake.*

*What are the actual side effects of the Pill? Most commonly mentioned are nausea, weight gain, and the danger of thrombophlebitis* [translator's note: *inflammation of a vein with a clot formation*]. *It is fair to say, though, that Pill-related ailments are very difficult to ascertain. They can only come to light through epidemiological studies. But in this case the margin of error is very large. Since the Pill is not the type of drug that is administered in hospitals, accurate long-term studies on its use and side effects cannot be carried out. We have to make do with doctors' reports, which are sporadic and often unclear about cause-and-effect relation. In any case, some side effects appear to be more prevalent in the developing countries. In Egypt, for example, there seems to be a fairly high incidence of liver disorders, while in Iran galactopoiesis* [translator's note: *formation and secretion of milk*] *is often associated with the use of oral contraceptives.*

*A comprehensive study on the subject has recently been published by the English journal* The Lancet.[6] *It concerns a collection of 46,000 women whose health had been accurately monitored since 1968 that had taken the Pill on a regular or occasional basis for varying periods of time. The study shows that the incidence of uterine cancer among Pill users is lower than among women who do not take oral contraceptives. This is probably attributable to the fact that the former undergo more frequent gynecological examinations. On the other hand, the incidence of cardiovascular disease is decidedly higher among Pill users; in the test group the death rate for cardiac arrest or stroke is 40% higher than in the control group.*

*Mention should now be made of one aspect of the problem that makes the use of contraceptives quite a different matter from, say, the use of fertilizers. With regard to the latter, the situation is clear-cut. On the one hand, there is a real*

*need for fertilizers in order to increase agricultural production; industry, pri-*
*vate or state owned, tries to meet this need, the one for profit, the other by*
*government policy. On the other hand, there are concerned people who advo-*
*cate a more rational use of fertilizers to prevent damage to the environment.*
*Nobody, however, would dream of asserting that the use of fertilizers is a mortal*
*sin, or that it is meant to promote some sort of genocide in the developing*
*countries. Consequently, the fertilizer controversy is largely carried out on a*
*technical level and is not greatly distorted by partisan statements or groundless*
*accusations. In the case of contraceptives it is just the opposite. In the Catholic*
*countries (600–700 million people), religious opposition to the Pill is un-*
*yielding and recently has been reiterated by the papal encyclical* Humanae
vitae. *Although a special committee appointed by Paul VI went as far as to*
*pronounce in favor of contraception, the Pope chose to adopt the opposing*
*minority view.*

*In all these countries every scientific finding, every piece of news or gossip*
*tending to prove that the Pill is unsafe is usually given prominent publicity. To*
*safeguard the principle that contraception is ethically inadmissible, the means*
*that make contraception possible are attacked, over and over again, as serious*
*health hazards.*[j] *In addition, many developing countries do not agree at all on*
*the need to curb their birth rates. At the World Population Conference held in*
*Bucharest in August 1974, the question of birth control was closely tied to the*
*problem of national growth and the economic relations between developing and*
*industrialized nations.*

*Many of the poorer countries tend to view the attempts of the affluent nations*
*to curb birth rates as a dangerous course "that threatens to aggravate their*
*underdevelopment, particularly because it is the low population density that in*
*some cases constitutes an obstacle to growth." This remark is taken from an*
*isssue of the magazine* Concilium *entirely devoted to the quality of life.*[7] *In the*
*same article the author notes that "the encyclical* Humanae vitae *has been*
*interpreted in Latin America as an antiimperialist manifesto."*[k] *He goes on to*
*summarize the situation as follows: "When the rich want to teach the poor how*
*to produce fewer children, the poor always wonder whether this is due to a*
*desire to alleviate their poverty or to the wish to limit the damage caused by their*
*proliferation."*

*Analogous sentiments are expressed by minority groups. As Djerassi writes in*
*an article in a 1969 issue of* Science *dealing with prospects for better oral*
*contraceptives and ways of making them available to mankind, "Some of the*
*economically deprived black inhabitants of our urban ghettos attribute geno-*
*cidal motives to family-planning programs in their areas."*[5] *Djerassi noted fur-*
*ther, in a 1966 issue of* Science, *that to this Babel of discordant voices the*
*press adds a touch of sensationalism by reporting some evidence linking the use*

*of oral contraceptives with thrombophlebitis under the heading "The Pill kills."*[5]

It is undeniable that current oral contraceptives fall short of perfection in many respects. They may pose a risk to the very young. Many authorities agree on this point. At the International Seminar on Fertility Control and Contraception held at Genoa in March 1977, Mrs. Kustrin of the Lubiana Institute for Family Planning stated, "A premature hormonal contraception can be dangerous and should not be initiated until at least 3 years after the onset of menstruation. On the other hand, the introduction of the intrauterine device can also be harmful to a not fully developed uterus. We recommend first the use of the diaphragm and the condom, then the intrauterine device, and the Pill only to problem adolescents, after an abortion."[8]

Note should also be made of a very serious incident caused by the clinical use of a synthetic estrogen called stilbestrol. Since this hormone can pass through the placenta, a significant fraction of the daughters of women who had taken stilbestrol developed cancer of the reproductive system in their middle teens. The great delay with which the long-term effects of stilbestrol were discovered is a measure of the difficulty of ascertaining this kind of danger in humans and of the extraordinary number of unknowns associated with the protracted use of apparently harmless substances. It goes without saying that the oral contraceptives currently on the market do not contain stilbestrol.

Aside from the question of side effects, admittedly the Pill is not the best birth control method in the poorer countries with low educational levels that need it most, at least in the judgment of most Americans and Europeans. It is relatively expensive ($1.25–2.50 per month) and requires a certain level of education. It would clearly be desirable to find a better method.

In the article by Djerassi mentioned earlier, a lucid and whimsical picture of the paradoxical aspects of the situation is painted. The article, entitled "Birth control 1984" in an obvious reference to Orwell's celebrated book, examines the possibility of finding the best possible means of contraception. According to Djerassi, the ideal contraceptive with the greatest chance of success even among poorly educated people is one that in a single dose will cause permanent but reversible sterility, that is, which can be annulled at will by the administration of a second drug that restores fertility. Second best are pills or any other means of contraception, in whatever way administered, that afford long-lasting effectiveness. Such products would make an appreciable contribution to the problem of birth control, although they would not solve it.

To achieve something close to what the author envisages, quite a few conditions must be met simultaneously. First of all, a formidable scientific problem must be solved. To develop a new contraceptive one needs to isolate or synthesize a large number of products, subject them to rigorous pharmacological

*screening, select the most promising, and then test them for short-, medium-,
and long-range effectiveness and toxicity. Clearly, an effort of this magnitude
cannot be borne by small research groups. It must be borne by large organizations, and there are not many that can afford even this initial step. Moreover, if
the ideal drug has to be structurally related to the substances that act on the
complex process of conception and in the very early stages of pregnancy, it must
be, to a first approximation, a steroid compound, a peptide hormone, a pros-
taglandin derivative, or one of their antagonists. There are very few organizations in the world that can afford competence in all three chemistries—few are
competent in one of them. Finally, the new drug will require very lengthy experimentation at all stages of research on higher and relatively expensive animals, such as dogs or monkeys, and with rigorous controls for a number of
successive generations so as to ascertain possible genetic consequences.*

*All of this costs an immense amount of money and requires highly specialized
personnel. To give an idea of the length of time involved, suffice it to say that
the Food and Drug Administration has imposed very strict regulations requiring that every new contraceptive be tested on monkeys for a period of 10 years.
(Note that these rules apply to* new *contraceptives. The products currently in
use have not undergone this type of testing, yet they have not been taken off the
market. This criterion, decidedly questionable, could be irreverently described
as an international version of the Neapolitan saying "Those who have, have.")*

*Clearly, an organization capable of developing new contraceptives can only
be found in the United States or Europe. In these countries, traditional methods
are already being used fairly successfully; consequently, what is to be expected
there from advanced research in birth control is an improvement in quality and
mode of use rather than the solution of the problem, which has largely been
achieved.*

*Once the initial stage of chemical and biological research has been com-
pleted, the new drug must be tested on humans. This is the last and perhaps
most difficult condition that has to be met. Human experimentation requires not
only time (7–8 years) but a certain degree of freedom. Recall now that the first
oral contraceptives were tested on Puerto Rican women and that it was only
after the encouraging results obtained, shall we say,* in corpore vili, *that the
Pill was introduced in the United States and Europe. This type of colonialist
practice, which allows for freedom of action and a modicum of secrecy, is completely out of the question today. Thus the developers of new birth-control drugs
will be faced with some unpleasant prospects, such as the need to test these drugs
on North American or European women and the sad but inescapable possibility
of having to interrupt a pregnancy occurring during their administration. And
of course there will be the related legal questions.*

*Even assuming that all this can be done, a minimum of 17–18 years will
have to elapse between the decision to initiate research and the time the product*

*receives the necessary authorization for marketing by health agencies such as the FDA. Development costs are bound to be very high. According to Djerassi's rough estimate, a total of at least $18 million (1965 dollars) is needed to market a new, improved, and easier-to-use contraceptive.*

*Given all this, what are the chances that this goal can be reached before world population doubles? To put it another way, can we close the barn door before all the horses have escaped? The outlook is not very encouraging. To begin with, there are very few pharmaceutical firms that can embark on a 20-year, $20-million research program. Furthermore, if there are few now, there will soon be even fewer, considering the commercial risk involved. Since a patent is valid for 25 years, and "possible" drugs must be identified and patented when research begins, the developer of the perfect product may be able to exploit its commercial value for a very limited time only. The current legislation, which provides that a 20-year research project be protected by a patent that expires in 25 years and in effect is commercially profitable for only a little over 5 years, does not seem too realistic.*

*In sum, we are in a no-win situation. The countries that need a new type of contraceptive, easier to use and therefore more suited to large-scale dissemination, do not have the means of developing it, while those that can carry out the necessary research are forced by their health regulations to make do with what they have.*

*The Current Situation*

Coming now to a final benefits-versus-risks evaluation, I think we may conclude that, on balance, chemistry's contribution to man's health has been positive. Some incurable diseases have been conquered; sexual habits and public mores have become more flexible; preventive medicine has made great strides. Research organizations and industrial groups have achieved a deeper understanding of human physiology and biochemistry, and this in turn has led to the discovery of new drugs. Part of industry's profit has been invested in research devoted not only to the development of new drugs but to the study of the most fundamental processes of life. There are all the conditions for even greater achievements. On the other hand, health care needs to be rationalized, and constant efforts must be made to prevent a drug's benefits from outweighing its potentially serious side effects, particularly its genetic effects.

As far as the development of new products is concerned, the current period is characterized by deep uncertainty. According to a study by Alfred Burger,[9] a highly qualified expert in the field of chemical applications to medicine, in the 10-year period from 1953 to 1963 about 50 new active agents were introduced every year into the US drug market. In the 1964–1974 decade the annual average fell to 16. The same decline was registered in patents of drugs originating from new applications of known active agents; over the two decades the number fell from 300 to 90. Although this trend is most obvious in the United States, it is observable to a greater or lesser degree in all technologically advanced countries. Is it a crisis in medicine, a slow-down in research, or the difficulty of obtaining licenses? Most likely, the decline in the development of new drugs is due to a combination of reasons.

The problem, as always, is to expand the physician's armory without introducing harmful substances into the health market. Maintaining this balance has never been easy. On the one hand, the manufacturers try to sell as much as possible and many physicians are apt to prescribe—and many patients are apt to consume—an excessive amount of drugs; on the other, health officials and responsible agencies attempt to impose strict limitations on the licensing of drugs that are not completely free from side effects.

The intervention of the regulatory agencies has been necessitated

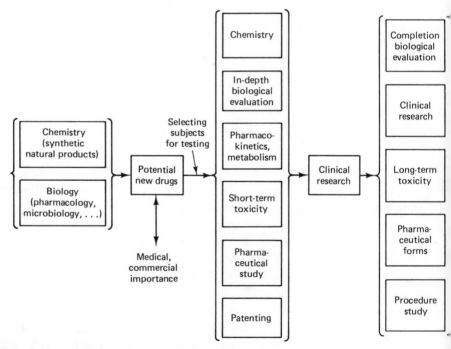

**Figure 3.1**
Schematic illustration of the stages of discovery and development of a drug according to current concepts.

by the rapid growth of the pharmaceutical industry, which in a very short time has replaced "galenicals" (infusions, vegetable extracts, and so on) with industrially produced drugs. Characteristically, it has always taken a tragedy to make people realize the need for stricter regulations. Consider what happened in the United States, for example. The Food and Drug Act of 1906, which set some restrictions on the free trade of galenicals and drugs, was amended in a restrictive sense only in 1938, after a small pharmaceutical firm put on the market a sulfonamide containing the solvent diethylene glycol from which some 80 people died of poisoning. The restrictions imposed in 1938 were reinforced in 1962 in the wake of the emotional reaction to the tragedy caused by thalidomide, the sedative responsible for the malformation of many European infants.

The current situation has contradictory and in some ways paradoxical aspects. A number of regulations have been introduced that set specific guidelines for the toxicological and pharmacological research required before a new drug can be approved for clinical use. To comply with these norms a company needs a solid technical organization and vast financial resources. Figure 3.1 illustrates the steps

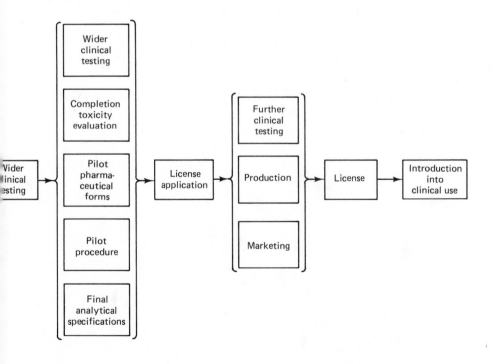

in the discovery and development of a new drug according to current requirements.

The net effect of all this has been to reduce drastically the number of firms that can develop new drugs and to limit them to a few that are very strong, and generally multinational. The cost of introducing a new drug has been estimated at about $10 million (pre-1973 dollars), a large fraction of which is earmarked for toxicopharmacological studies. "Therefore," as Bruger writes, "the pharmaceutical industry as a rule can study and develop only drugs with an adequate market potential: drugs for diseases dangerous in developing countries or for those that occur only rarely remain of only academic interest. Even if such drugs were discovered, they could not be carried through clinical trials by the pharmaceutical industry because of the expense of these trials. . . ."[9]

Long delays in obtaining licenses are quite common, at least in the United States. A survey conducted by Louis Lasagna and William Wardell of the University of Rochester shows that none of the products approved for human experimentation in 1968 had been licensed as of August 1974. According to Burger, this is a dangerous situation since it does not allow new, safer—although not perfect—drugs to replace products that were approved previous to the newer restrictions. In the view of many experts (including Djerassi), a good portion of the drugs sold today would not pass the most recent toxicopharmacological examination.

We are clearly in a period of readjustment and transition. Overcoming these difficulties in a positive, productive manner constitutes the current challenge to medical science.

# 4

## Chemistry, Hygiene, and Cosmetics

In no field is the contact between chemistry and the individual as direct as in cosmetology, a term used here to denote the practice of personal hygiene and beautification. Cosmetics are an item in constant expansion in the budget of the developed nations. Even in the least civilized country each person uses daily at least one or two products for the care of the body; several people use as many as a dozen, if not every day, at least in a short space of time. According to fairly reliable estimates, in 1970 Americans spent about $10 billion in toilet articles, at barber shops, and in beauty salons. With regard to Europe, 17,000 tons of shaving creams and shampoos were imported in 1974 at a cost of $26 million and 23,000 tons were exported with a value of $40 million. Soap imports (including medicated soaps, which are generally based on Marseille soap with antiseptic or medicating agents) amounted to 47,000 tons, or $59 million; exports, to 101,000 tons, or $118 million. Germany and France are the leading producers of perfumes in Europe. In 1974 perfume production in each country exceeded $1.1 billion, compared to Italy's $500 million. Perfume exports are one of the strongest points in France's balance of trade. In 1977 Italians spent about $1,500 million for toiletries, which represents a 20% increase with respect to the previous year. Exports in this sector are low, amounting only to 25% of production.

The socioeconomic implications of the cosmetic industry are not as easy to define as for other industries since they are largely tied to psychological motivations and evaluations. There is no question, however, that it is flourishing. This is a typical case in which the value of a product depends on a successful formula rather than on the cost of the materials, which are generally inexpensive. The intrinsic cost of a soap, cream, or perfume is relatively low, as evidenced by the fact that when a skin care product is sold as a medicinal remedy, and is therefore subject to price controls, it costs much less than when it is sold as a cosmetic.

## The Scientific Approach to Cosmetology

Cosmetics have come a long way from the beauty aids of the ancient Egyptians to the products sold today by sophisticated American or European houses. Until a few decades ago only natural substances were used for the care of the body. Advances in chemistry and a better understanding of the biochemical processes associated with skin physiology, dental health, and hair structure have put the manufacture of cosmetics on a scientific basis. Even in this field, so highly influenced by personal taste and psychological reactions, efforts are made to use a more rational approach.

Let us consider first the complex problem of hair care. We are all terribly preoccupied with our hair. When we have it we do all we can to keep it; when it is gone, we mourn its passing. We wash it, cut it, dye it, bleach it, streak it, wave it, and condition it; we constantly change haircut and hairdo, by hot or cold processes. Not including barber and hairdresser costs, in 1970 Americans spent over $1.2 billion for shampoos, dyes, hair sprays, permanent-wave preparations, and the like. Italians, on their part, spent $170 million in 1976 for hair and scalp care.

One of the key developments in hairdressing was the introduction about 1930 of the cold wave, which paved the way for the modern home permanent that does not require any elaborate equipment. We own this advance to scientists of the Rockefeller Institute who demonstrated that the sulfide bonds that give proteins their spatial structure can be split at low temperature and slightly alkaline pH by the action of sulfides or mercaptans. This observation created a fast-growing industry whose sales in the United States alone are estimated at more than $100 million. Commonly used for cold permanent waving are solutions based on ammonium and sodium salts of thioglycollic acid. Depilatory creams and lotions work on the same principle. The thioglycolates or sulfides contained in such preparations soften and loosen unwanted hair, which is then easily removed.

Chemistry has also made a contribution to the problem of holding hair in place for a number of hours, even when elaborately set. Hair sprays have allowed women to do away with the arsenal of pins, needles, and combs that our mothers used in order to keep their hairdos in precarious balance. At the same time they have opened new avenues for the use of plastic materials. The countless hair sprays currently marketed in aerosol cans (which use a combination of fluorocarbons as propellant) are lacquer preparations that form a film on the hair. They are based on a variety of active ingredients, such as polyvinylpyrrolidone, its copolymers with vinyl acetate, copolymers of maleic anhydride and vinyl acetate, and terpolymers of vinyl monomers containing ester or carboxyl groups.

Hair coloring is another field to which chemistry has made a large contribution. Annual sales of hair dyes amount to more than $270 million in the United States alone. Again we find a great variety of compounds, generally based on *p*-amino phenol or *p*-phenylene-diamine. Hair colors are often the same dyes used in the textile industry, and their application requires a series of preliminary treatments involving the use of various chemicals, such as oxidants, alkalizers and bleaching agents.

The application of hair-care products may be considered,within certain limits, as an "external use," in the sense that any contact between the skin and the chemical is limited and incidental or due to improper use of the preparation. Many cosmetics, on the other hand, involve direct contact. Such is the case for skin-care products such as nourishing, moisturizing, lubricating, and cleansing creams, all of which are usually applied in massive doses, repeatedly, and for long periods of time and are absorbed through the skin. A good example is the suntan lotion (oil or cream) that is applied to the skin to promote tanning and prevent sunburn. Although the use of such preparations is very much a matter of personal taste, it generally involves protracted applications over periods ranging from a week to months, which results in the absorption of considerable amounts of chemicals.

Suntan lotions contain a variety of substances. In the early days of the cosmetic industry, until about 1945, they were preparations whose origin could be traced to popular traditions and uses. Careful research was subsequently carried out to determine the relation between the wavelength of the incoming sunlight and the processes of tanning and sunburning. It was found that the skin is affected solely by radiation at wavelengths from 2,970 to 3,300 angstroms [*translator's note:* by definition, a unit of wavelength 1/10,000,000,000 of a meter]. Both effects—tanning and burning—proceed in parallel at wavelengths from 2,970 to 3,150 angstroms, while radiation from 3,150 to 3,300 angstroms causes tanning but not burning.[1] Thus the ideal preparation is one that screens out radiation from 2,970 to 3,150 angstroms but lets through that from 3,150 to 3,300 angstroms. In the past the most commonly used substances were derivatives of salicylic, anthranilic, and para-aminobenzoic acids, followed later by 2-

ethoxyethyl-*p*methoxycinnammate. More recently, dihydroxyacetone
and glyceraldehyde have been successfully used in aerosol cans. Annual sales of suntan lotions in the United States range as high as
$80–90 million.

Deodorants and toothpastes are two additional products that come
into direct contact with the organism. In this case, too, commercial
production was limited at first to traditional ingredients and methods;
and, again, we owe to a more rational approach the development of
new and more effective, although not necessarily safer, preparations.
In the case of deodorants, for example, the zinc oxide formulations
commonly used in the first half of the century have been replaced by
compounds containing strong antibacterial agents. Research has
shown that the occurrence of unpleasant body odor is almost always
due to localized bacterial action. For many years hexachlorophene
was the antibacterial agent par excellence, but its toxic effects have
prompted the FDA to restrict its use. The history of hexachlorophene
is somewhat complex. For a long time it was considered a harmless
substance. Then some years ago a talcum powder sold in France, to
which a much higher amount of hexachlorophene than normal had
been added by mistake, caused the death of several infants. The
tragedy was attributed to this bactericide, which was subsequently
banned. It should be noted, however, that in general it is not quite
clear whether responsibility for certain cases of poisoning should be
ascribed to hexachlorophene per se, since research has shown that it
may contain nearly the same amount of dioxin found in the weed
killer 2,4,5-T.[2] Current deodorants contain other bactericides.

Dental care is one of the daily concerns of civilized man. Modern
toothpastes have the double function of cleaning teeth and helping to
prevent dental caries. Their considerable success in the latter respect
is due to the use of chemicals, such as stannous fluoride and sodium
monofluorophosphate, that have proved quite effective in the prevention of tooth decay.

*Health Considerations*

To this brief summary of the state of the art we should add some
cautionary remarks. In this field, apart from the economic and industrial aspects, health considerations should be given particular relevance. Consumer protection is surprisingly limited in the cosmetic
field, at least in a relative sense: In contrast to the great deal of attention given to drugs and food additives, there is an almost total indif-
ference with regard to soaps, perfumes, creams, make-up, hair dyes,

toothpastes, nail polish, and in general to all those preparations—often of an undisclosed nature—that everybody can freely buy and use in massive doses and for long periods of time.

These products are made by established firms as well as by cottage industries. They are sold in drugstores and supermarkets and even peddled door to door in unsealed containers of dubious origin. Their exact composition is never spelled out—not even in the best known, most sophisticated, and expensive products. At the same time such extravagant claims are made for their beneficial effects that, if true, they could only be achieved by substances of great biological potency (which therefore should be carefully controlled). The case of hexachlorophene is typical: One cannot help wonder how a substance that is often contaminated by dioxin impurities and that in large amounts can cause death could possibly find its way into talcum powder for infants.[a] Admittedly, the French tragedy was caused by an error in dosage, but the real cause was the prevailing carelessness in these matters.

The inconsistency is obvious and, unfortunately, widespread. Consider common toothpaste, for example. The amount of detergent, lead, and fluoride with which we happily rinse our mouths would never be allowed in industrial effluents by Italian water-safety regulations. It is a fact that while some sectors have been carefully regulated, the cosmetic field, which is much closer to man, has been almost totally neglected.

*Recent Initiatives*

Some steps have been taken to correct this situation. I shall mention two. The first originates with US cosmetic manufacturers, who have joined together in a program aimed at reviewing the safety of the ingredients used in their products. The project, called Cosmetics Ingredients Review, is financed by the Cosmetic, Toiletry and Fragrance Association, but will be carried out in complete freedom of action and independence from the manufacturers. The activities of the Cosmetic Ingredients Reviews are to be conducted by a number of experts chosen among academics, industrialists, government officials, and consumers. The FDA has been asked to name a representative, and the commission will be headed by Burger, who is a member of President Carter's committee of scientific advisors. A report will be prepared for each of the ingredients under scrutiny, and all information gathered by the commission will be made public. The cosmetic industry has understood the need to eliminate the risks connected with the use of its products as well as the desirability of sharing its knowledge, lest uncertainty should result in unfavorable regulations.

A second and more far-reaching measure has been taken within the European Economic Community. Some years ago the Community took note of the fact that cosmetics had achieved widespread diffusion and that all kinds of preparations were being introduced into the market. This situation, common to all member countries, clearly called for legislation that would safeguard the health of the consumer. In 1976 the Council of the EEC issued a directive (76/768/EEC, 27 July, 1976) to be implemented in the member countries that defines cosmetics[b] and gives their main distinguishing characteristics. It also lists some substance that can safely be used in cosmetic preparations, proposes restrictions on the type of materials to be used as containers, and gives guidelines on packaging, labeling, advertising, expiration dates, and so on. The implementation of this directive would also impose some regulations on the manufacturers, including the hiring of technical personnel as part of the work force and making sure that plants and equipment meet certain standards. The EEC directive will be shortly transformed into law in the member countries and should mark a significant improvement in consumer protection. In Italy the passage of a cosmetics regulatory law has often been reported as imminent.

# 5

---

## New Materials from Petrochemicals

### Plastics

Almost everything we touch is in some way connected with high-polymer plastics.[1] The fact that traditional materials such as wool, silk, wood, and metal have been wholly or in part replaced by plastics and other synthetic materials may inspire some refined souls with romantic regrets, but we must not forget that it has enabled less affluent people to own a multitude of objects of everyday use that they could not otherwise afford. In addition, plastics have provided alternatives for materials like wood whose large-scale use would cause serious ecological problems; obviously, to obtain enough wood to make all the things that are now made of plastic we would have to deplete the world's forest reserves. These positive aspects are unfortunately accompanied by negative ones, to which we shall shortly return.

The growth of the plastics industry has been spectacular in the last 30 years. At present, sales exceed $5 billion in the United States alone. In Europe the annual consumption of plastic materials amounts to more than 16 million tons, of which 4.7 million are used in Germany, 2.6 million in France, and 2.15 million in Great Britain and Italy. This success is due to the fact that the basic compounds—petrochemicals—are relatively abundant and inexpensive and can be turned with fairly straightforward procedures into high-quality, high-performance materials that are light, durable, and economical; about 90% of synthetic organic compounds originate from petroleum and natural gas; the remainder comes from carbon residues of coke plants, animal and vegetable fats, resins, and vegetable seeds. In the specific field of synthetic resins, basic materials include ethylene, propylene, styrene, and benzene, all of which can be obtained in vast quantities during oil-refining processes.[2] The annual production of ethylene alone exceeds 20 million tons.

Plastics may be roughly divided into thermoplastic and thermosetting types. Thermoplastic materials are generally soluble and
characterized by their ability to remain plastic after numerous
heating treatments. They include polystyrene, polyvinyl chloride,
polyethylene, and polypropylene. Thermosetting plastics, once
formed at high temperatures, are insoluble and infusible and include
phenol and epoxy resins, and styrene-based polyesters.

Plastics have a wide range of applications, ranging from insulating
material to household objects (telephones, doorknobs, switches, cut-
lery), foils, textiles (washable as well as disposable), syringes, paper,
ropes, glues, unbreakable bottles, and underwater cables. More recently, plastics have found wide application also in the biomedical
field, notably in the production of artificial organs and prosthetics and
by providing techniques for the slow release of drugs within the body.
(The technique of encapsulating a drug in polymeric materials so as to
allow its gradual release has the advantage of keeping it at the appropriate level in the organism for a long period of time and avoiding the
various problems associated with its frequent administration.)

Plastics have two types of drawbacks: one is the difficulty of dispos-
ing of them satisfactorily, and the other is the toxicity of many of the
basic materials used in the mass production of polymers. The reason
why the disposal of plastic materials is such a serious problem is that
most of the high polymers are practically indestructible. This fact,
along with the low cost of plastic products and the poor civic sense of
the consumers, has caused the affluent nations to be deluged by plastic
garbage. Every recreation area is littered with plates and cups discarded by holiday crowds; plastic trash floats in lakes and down rivers;
our cities are covered with plastic bags, containers, and toys; our
dumps are swamped every day with hundreds of tons of plastic products that are almost impossible to get rid of. This type of pollution is
the result of a widespread phenomenon that everybody complains
about and everybody helps to create. And this is why it is so hard to
control. In recent years, however, public campaigns have started in
the industrial countries to limit the spread of plastic waste, which is
certainly an encouraging development.

The response of the manufacturers has been twofold. At first they
turned to the development of degradable polymeric materials. For
industry, this attitude represented a complete turnabout since its aim
had always been to create long-lasting materials that could withstand
weather, wear, and other environmental factors. The problem now
was to manufacture materials with a finite life, that is, materials that
would perform the function for which they had been designed but,

once discarded, would be broken down by the action of sun, air, and microorganisms. As a result, new consideration was given to all those monomers that in the past had been carefully avoided because of their undesirable characteristic—now a virtue—of imparting to the polymers a certain degree of degradability.

The energy crisis, coupled with the scarcity of raw materials, has caused a shift in priorities from waste disposal to recycling, and research has been initiated on a number of processes that would enable us to produce raw materials from waste products. This change of attitude can also be observed in the plastics field. Many industrial groups are now studying ways of recycling plastic materials as such and of producing basic materials for the petrochemical industry from polyethylene, polystyrene, rubber, and so on. By a process developed in Germany, polyethylene can furnish as much as 25% (by weight) ethylene, 16% methane, and 12% benzene. Polystyrene can yield up to 75% styrene.[3]

The second drawback of plastics is far more worrisome. Every year petrochemistry introduces into circulation millions of tons of basic materials such as ethylene, acetylene, benzene, styrene, vinyl chloride, and vinylidene chloride. These compounds form the basis of industrial organic chemistry and some of them are essential for the production of plastic materials. In recent years, because of the new awareness of the problems caused by the diffusion of chemicals in the environment, many of these basic materials have been reexamined from the toxicological standpoint. The toxic effects of vinyl chloride are practically old news (see chapter 1). Many other substances give cause for concern, notably benzene, which is extensively used in industry as well as in research laboratories, styrene, acrylonitrile, and vinylidene chloride. Naturally, concern is more serious, the more widespread the product.

*Synthetic Textiles*

The development of man-made fibers has truly revolutionized the textile industry. In the United States, for example, since the late 1960s the production of man-made fibers has exceeded that of cotton and wool, while silk has practically disappeared from the market. The great abundance of synthetic fibers (nylon, polyacrylic, polyester, rayon, acetate) has thoroughly changed the way people dress. Everyone's wardrobe has become much richer. There is a greater variety of summer and winter clothes than at any other time in the past, and buying new clothes has become more a question of fashion than of real need. Draperies and carpets are more affordable. Fabrics are easier to care for, more durable, and available in an almost infinite choice of colors. In the less affluent countries buying clothes is no longer a prerogative of the rich. From the economic point of view, recent statistics show that the average person in the West currently spends 9% of his income for better and more plentiful fabrics than he could obtain 30 years ago by spending 12% of his income.

Thanks to chemistry, the textile industry has been able to avail itself of numerous technical innovations. In addition to the discovery and mass production of many new materials, great strides have been made in the development of processes that make fibers wrinkleproof, nonflammable, less transparent, dirt resistant, and easier to dye. Cotton is an example of a natural material that loses its original crease upon washing. This is due to the fact that the weak forces that bind the cellulose molecules tend to weaken in water, so that the molecules are rearranged in new spatial configurations. Fabrics treated with special chemical agents become more resistant to laundering, even if done by the rather drastic methods of modern washing machines. The permanent-press fabrics currently on the market consist of cotton, nylon, and blends of natural and synthetic fibers. The growing use of washing machines has created problems also for wool garments. The combination of heat, chemical cleaning agents, and mechanical vibrations used in automatic washers causes wool to unravel and felt. This problem is obviated by treating the surface of the cloth with synthetic resins, along with a slight chloridation. Chloridation processes, as well as treatments with formaldehyde or chlorosulfonic agents, are extensively used to make wool cloth more resistant to microorganisms. (Mothproofing of wool products was achieved for a while by the addition of dieldrin, but this practice has fallen into disuse because of the marked toxicity of this insecticide.)

Chemistry also plays a part in the waterproofing of garments. In this case, too, difficulties arise from the frequent use of automatic washers. In effect, the advent of the washing machine revolutionized waterproofing treatments. Until the 1950s waterproofing agents were based on aluminum or zirconium hydroxides. They were subsequently supplanted by quaternary pyridine compounds combined with fluoride derivatives. At present, most commonly used are mixtures of fluoropolymers, pyridine derivatives, and silicone compounds. It is worth mentioning that a great deal of research in this field was carried out for military purposes and that the products we use today are the result of these studies.

An important aspect of textile chemistry is the use of the so-called fire retardants,[4] that is, substances that make fabrics resistant to fire. To understand the importance of these chemicals, suffice it to say that according to *Chemistry and Economy,* fire is the third greatest cause of accidental deaths after car accidents and falls. Combustible fabrics have been found to play a leading role in most fatal fires. Chemistry has contributed to the solution of this problem with the development of phosphochlorinated compounds that render fabrics incombustible. In 1970 consumption of fireproof fabrics amounted to 500,000 tons, and by the end of the decade it is expected to reach 2 or 3 million tons. In 1977 about 160,000 tons of fire-retardant agents were used in the United States alone.

Technological developments have made it possible to create synthetic fibers but also materials of an entirely new nature. A good example is "nonwoven" fabrics. Although such fibers are not woven but held together with special techniques, these fabrics look woven. They are not too strong, but they are very inexpensive and are used to make disposable clothing and linens, particularly for hospital use. Smocks, sheets, towels, and bandages made of nonwoven fabrics are used once and then thrown away, thus removing the need to wash and sterilize these items frequently and thereby reducing the danger of infection.

Along with these advantages, synthetic fabrics have a number of disadvantages. To begin with, man-made fibers are not always compatible with the human organism. Nylon, for one, is known to cause allergic reactions. In addition, most of the chemicals used to make fabrics waterproof, crease resistant or incombustible are toxic to a certain degree. Again, it is a question of choice. If we want the convenience of washing machines and permanent-press shirts, we must accept chemical treatments that are not always safe. Fire retardants are also under investigation, and restrictions on their use have been advocated by many. The decision will ultimately rest with the regulatory health agencies and, once again, it will depend on a careful risks-versus-benefits evaluation of the products, in this case their potential health hazards versus the benefits of preventing the terrible consequences of fire and burns.

### Synthetic Rubber

World demand has far outstripped the supply of natural rubber. In this field, too, chemistry has helped to meet world needs by developing synthetic raw materials as well as additives that improve their performance. The manufacture of synthetic rubber is closely related to the chemistry of high polymers. Synthetic rubber is produced in huge quantities from petrochemical derivatives by processes that require the use of sophisticated catalysts. More than 3 million tons of synthetic rubber is produced annually in the United States alone; styrene-butadiene rubber is the first in order of quantity, followed by polybutadiene, neoprene, butyl rubber, and polyisoprene.

A large number of chemical additives are employed to improve the performance of rubber products. Vulcanization processes, which are necessary to give tires strength and durability, are accelerated by zinc salts, zinc oxide, and thiazole derivatives. The life of a tire is extraordinarily prolonged by the addition of lamp black (amorphous carbon), which is currently produced in millions of tons.

The production and use of synthetic rubber create considerable health hazards. Some of the basic monomers are intrinsically unsafe; styrene, for example, is suspected of being carcinogenic. The finished products, moreover, always contain residues of the catalysts employed in the polymerization process, which are generally metals with toxic properties. In addition, the use and destruction of tires contribute to environmental pollution by disseminating a large quantity and variety of chemical additives. It is symptomatic that by the side of the highways, besides high concentrations of lead (originating from the tetraethyllead contained in gasoline) one also finds a certain amount of zinc, which comes from the zinc oxide in tires, and of cadmium, which accompanies zinc as an impurity.

## Coatings and Glues

Another field in which chemical advances have benefited both the community and the individual is that of protective finishes. Coating materials have countless applications and enable us to protect objects, houses, pipes, and boats against weather, dampness, pollution, and salt-and freshwater damage. They prevent corrosion, rust, and deterioration and at the same time enhance appearances with a constantly growing palette of colors. We all use paints, lacquers, and varnishes directly or indirectly. Sales of coating materials are estimated at 5% of the entire chemical market.

With regard to composition, coatings are based on a great variety of compounds. Nitrocellulose lacquers have given way to latex paints made from urea-formaldehyde or melamine-formaldehyde resins, styrene-butadiene emulsions, and epoxy and acrylic resins. Pigments based on white lead have been supplanted by titanium dioxide, which is much brighter, cheaper, and less toxic. Sophisticated coating materials have been developed that control the diffusion of water, oxygen, or light, act as lubricants, inhibit mildew, and permit easy cleaning. (We are all familiar with "no-stick" pots and pans.)

Other products we almost take for granted are adhesive tapes and miracle glues that bear enormous weights or bind together metallic surfaces in a matter of minutes or even seconds. Again we are dealing with epoxy resins and special rubbers, all of which are accurately selected for maximum performance with a minimum of bulk, weight, and cost.

## Soaps and Detergents

In the 1930s the soap and detergent industries experienced a spectacular growth. This expansion coincided with the advent of the petrochemical industry and the replacement of agrochemical raw materials with petroleum derivatives. Another factor that was at the same time a cause and an effect of the drastic change that took place in the detergent industry was the development of the automatic washing machine and its use on a world scale.

Up to 1930 the raw materials of the detergent industry were fatty acids originating for the most part from the hydrolysis, or *saponification,* of animal and vegetable fats. The soaps obtained by this process were excellent except for a certain incompatibility with hard water. In the presence of calcium and magnesium ions, soaps of fatty origin tend to precipitate, separate from the liquid phase, and form deposits on containers. This characteristic made them unsuited to automatic washers because drain pipes tended to become clogged by detergent precipitates.

The first detergents of petrochemical origin were alkylbenzene sulfonates. The chemical properties of these compounds make them particularly suited for use in washing machines, as well as in toiletries and in some industrial processes requiring the use of detergents. Manufacturers have since developed a wide range of specially formulated detergents designed to meet the most disparate needs with extraordinary efficiency. On the one hand, the surfactants have been improved by varying the length or ramification of the alkyl chain; on the other, various ingredients have been added to the basic detergents in order to speed up the cleaning action (as required by automatic washers), prevent felting, and make clothes whiter and brighter. Over the years advertisment has entered the profitable detergent market with massive campaigns aimed at an increasingly consumer-oriented society. Since modern household detergents are basically all the same from the point of view of performance, manufacturers compete on the basis of extravagant claims and tricks that catch the consumer's fancy. One cleaner is said to "make the window disappear", another "to make clothes whiter than white." In other words, advertisement has turned the detergent market into a mixture of real needs and catchy slogans.

The phosphates (pyrophosphates and tripolyphosphates) have long been considered the second most important ingredient of detergent preparations because of their ability to prevent the formation of calcium and magnesium deposits in washing machines and to increase detergent action, particularly for cotton fabrics. Most detergents also contain small amounts of carboxymethyl cellulose which prevents the gradual yellowing of cotton fabrics. Fluorescent whitening agents are added to restore whiteness and brightness. Disinfectants are sometimes added to soaps and detergents to reduce microbial contamination. Hexachlorophene was extensively used for this purpose until banned. Quaternary ammonium salts have successfully replaced the hypochlorites, which also had a marked corrosive action. In the 1960s detergents containing proteolytic enzymes enjoyed great popularity because of their effectiveness in removing soil, but the use of "biological" detergents was discontinued when they were found to cause allergic reactions and lesions on the hands.

Detergents also have various industrial applications; they are employed, for example, in levigating surfaces that must undergo special treatments and in some chemical processes requiring the use of emulsifiers, surfactants, and so on.

The history of detergents has been marked by a number of contradictions, revisions, and technical advances aimed at meeting changing needs. In the 1960s the general public suddenly awakened to the fact that many streams, lakes, and seas were heavily polluted by detergents, although this phenomenon had long been a part of our daily lives. Such high concentrations were due partly to the growing use of detergents and partly to their low degradability in the environment. Knowing the biological processes through which microorganisms break down organic substances, the cause of the problem could be traced to the branched alkyl chain contained in detergents. Microorganisms cannot readily break down this type of structure, while they can quickly metabolize linear alkyl chains. This finding was promptly put to use by the manufacturers, who in a short time and without excessive cost increases were able to develop detergents with a linear alkyl structure. It also prompted legislative action; practically all countries banned the use of nonbiodegradable detergents. As of 1

January 1971, Italian law requires all detergents to be at least 80% biodegradable.

Another problem associated with the use of detergents stems not so much from the cleaning agents themselves as from the polyphosphates that are added to most laundry detergents. Phosphates, along with nitrogen compounds, are one of the nutrients that stimulate plant growth. The nitrates and phosphates contained in fertilizers as well as the polyphosphates from detergents eventually find their way into many bodies of water, and this excess of nutrients cause the phenomenon known as eutrophication, that is, the massive uncontrolled and largely uncontrollable proliferation of algae. Eutrophication is particularly evident in lakes like Lake Erie and enclosed seas, such as the Adriatic and the lagoons around Venice, where water is renewed very slowly. Because of the seriousness of this problem, detergent manufacturers are reviewing their formulations in order to decrease the proportion of phosphates or replace them with substances just as effective from the practical point of view but lacking their undesirable side effects.

*Fluorocarbons*

Fluorocarbons are fluorinated hydrocarbon derivatives with a small number of carbon atoms, chiefly used as coolants in refrigeration and air-conditioning equipment and as propellants in aerosol cans.[5] Fluorocarbons are liquid at room temperature and moderate pressure, nonflammable, generally nontoxic, and can be easily and inexpensively produced from halogens and methane. Because of such desirable properties, they have replaced ammonia and sulfur dioxide in refrigerating units and carbon dioxide in aerosol spray cans. The best-known compounds in the Freon group are Freon 11 (trichlorofluoromethane, $CFCl_3$) and Freon 12 (dichlorodifluoromethane, $CF_2Cl_2$). In the 1958–1973 period, 2.5 million tons of Freon 11 and 3.8 million tons of Freon 12 were produced and sold worldwide. In 1973 US production amounted to 70,000 and 100,000 tons, respectively.

In June 1974 an article by M. J. Molina and F. S. Rowland[6] focused world attention on the potential dangers associated with the growing use of fluorocarbons. The two scientists pointed out that compounds like Freon 11 and Freon 12 may diffuse into the atmosphere without breaking down and eventually reach the ozone layer in the stratosphere. By reacting with the ozone they may cause an appreciable depletion of the layer itself. Should this happen, the consequences for mankind would be very serious. Ozone ($O_3$) is formed in the earth's upper atmosphere by the photochemical action of solar ultraviolet light. The ozone layer thus formed acts as a shield for life on earth by filtering out most of the harmful high-energy ultraviolet radiation from the sun. More precisely, the ozone layer absorbs most of the radiation between 2,800 and 3,200 angstroms, which is called erythematous radiation because it can cause erythemas, or burns, in people with sensitive skin. As shown by epidemiological studies, this radiation may have far more serious effects than sunburns. In particular, the incidence of skin cancer and the amount of erythematous radiation that reaches the soil appear to be related to latitude, which suggests a cause-and-effect relation between such radiation and skin cancer. Thus, the depletion of the ozone layer and the resulting increase in erythematous radiation may have adverse effects on man's health as well as on the climate.

Although many scientists have confirmed its general validity, this hypothesis is very difficult to prove. Direct measurements in the stratosphere are not easy; nor is it easy to evaluate variations in the composition of a gaseous medium. Furthermore, wide variations can be produced by perfectly natural causes, such as thermal inversions or updrafts. Consequently, attributing a variation to a specific cause is at best an uncertain proposition. Recent studies have also cast doubts on the basic assumption, namely, that fluorocarbons remain unaltered in the atmosphere. It has been suggested that they may break down to a certain extent and therefore never reach the stratosphere, or reach it only in limited amounts.[7]

The diffusion of fluorocarbons has been rapid and massive, whereas their interaction with the ozone in the stratosphere is assumed to be a slow process. This means that the study of this phenomenon will take a long time. Thus, although in theory the danger is real, there is no time to assess it. On the other hand, waiting to ban the use of fluorocarbons until we have more information means introducing more and more of them into the atmosphere, thereby making the danger more serious, if not catastrophic. At present both manufacturers and health agencies have taken a cautious attitude on the issue of fluorocarbons, and numerous limitations have been imposed on their use.[8]

# 6

## The Industry of Inorganic Compounds

### Basic Inorganic Compounds

The chemical industry produces an enormous amount of inorganic compounds. These may be divided into two groups: basic inorganic compounds and metals or metal derivatives. Basic inorganic compounds constitute approximately 20% of the total output of the chemical industry. They are produced partly for commercial purposes and partly to be used as reagents by the producers themselves. There is hardly any chemical production in which one of these compounds is not used to some extent and in some form.

Consider first sulfuric acid, which is one of the most important chemicals from the industrial point of view and is produced everywhere in very large quantities. Sulfuric acid is used in numerous chemical processes as a dehydrating agent, catalyst, oxidant, and solvent as well as acidifier. About 40% of the world output of sulfuric acid (over 100 million tons in 1975) is employed in the manufacture of phosphates, 10% in the production of ammonium sulfate, another 10% in petrochemical processes, and 20% in various other industrial activities.

Ammonia is quantitatively the second most important industrial product. It is produced in amounts of 40–50 million tons, chiefly by the Haber-Bosch and Fauser processes. The amount of nitrogen that industry converts to ammonia and other nitrogen compounds is approximately 80 million tons per year, while the amount "fixed"[1] by biological organisms is by a very rough estimate about 175 million tons per year. Beside being the source of nitrogenous fertilizers, ammonia is used to produce other compounds, such as nitric acid, and as a reagent in the manufacture of sodium carbonate.

Chlorine and caustic soda are also produced in large quantities (about 25 million tons each), and both have a very wide range of industrial applications. Caustic soda is essential to the production of fibers from cellulose, paper, soaps and detergents, phenols, and certain metals as well as to numerous other industrial processes. Chlorine is extensively used in the textile industry; in the manufacture of chlorinated insecticides, fluorocarbons, industrial solvents (such as methylene chloride, chloroform, Trieline), and plastics (e.g. polyvinyl chloride); as a bleaching agent in the paper industry; and as a disinfectant to purify water supplies. In addition, chlorine is the basic material for the preparation of compounds such as hydrochloric acid, which finds large use in the steel industry and in metal refining as an alternative to sulfuric acid; and hypochlorites, chlorites, and chlorates, which are employed in the manufacture of explosives, matches, weed killers, and so on.

Sodium carbonate is an essential material in the glass industry and various other industrial activities. World production of sodium carbonate is around 18 million tons.

## *Metals*[a]

As *Chemistry and Economy* remarks, "The strength of an industrial nation rests heavily on its capacity for making iron and steel. Nonferrous metals, concrete, wood, and plastics may compete with ferrous metals in many uses, but none is so fundamental to a country's production of goods and services." What makes steel so unique is the fact that it combines strength, hardness, malleability, and low cost. Some of the steel industry's largest customers are the automotive and construction industries, followed by machinery, containers, rail transportation and ship building, but there is no industry that does not require the use of steel in one way or another.

Aluminum, copper, lead, tin, nickel, zinc, and cadmium are the most common and most widely used nonferrous metals, but platinum, chromium, germanium, indium, thallium, and the lanthanides are also important from the commercial and industrial point of view. Chemistry often plays a major role in some stage of the manufacture of these metals. Chemical reagents are used extensively, and in special processes they are needed to separate a metal, or one of its ions, from a mixture in which it is present in low concentration. Since the richest mineral deposits are close to depletion, production is increasingly dependent upon more sophisticated refining techniques.

The world output of nonferrous metals is very large. In 1974, for example, the EEC produced about 2 million tons of aluminum, 1 million tons of both copper and lead, and 1.5 million tons of zinc. Italy

**Table 6.1**
World production of the principal nonferrous metals (in tons)

| Metal | 1973 | 1974 | 1975 | 1976 |
|---|---|---|---|---|
| Aluminum | 27,722,800 | 30,003,900 | 27,555,100 | 28,102,200 |
| Gold[a] | 3,335,000 | 3,141,000 | 3,010,000 | 3,078,000 |
| Cadmium | 17,504 | 17,326 | 15,523 | 17,021 |
| Cobalt[b] | 21,850 | 24,550 | 21,100 | 18,000 |
| Refined copper | 8,513,000 | 8,871,900 | 8,381,800 | 8,871,800 |
| Mercury[c] | 286,100 | 279,330 | 277,290 | 257,170 |
| Magnesium | 251,020 | 269,298 | 245,452 | 240,508 |
| Nickel | 656,916 | 718,612 | 709,057 | 751,093 |
| Lead | 4,197,500 | 4,255,500 | 4,021,300 | 4,104,900 |
| Antimony | 37,394 | 38,065 | 37,116 | 40,392 |
| Tin | 224,883 | 221,823 | 220,396 | 222,250 |
| Zinc | 5,831,800 | 5,993,700 | 5,485,900 | 5,815,700 |

Source: *Annuaire Statistique MINEMET*, Paris (1976).
a. In pounds.
b. Excluding China and the communist bloc.
c. In 76-pound bottles.

produces substantial quantities of these metals from minerals that are for the most part imported. Table 6.1 lists the production figures for the principal nonferrous metals for the years 1973–1976.

Nonferrous metals and their derivatives have a wide range of applications and an equally wide diffusion. Owing to its ductility and high electric conductivity, copper is the ideal material for the manufacture of electric cables, wires, switches, and electromechanical equipment. Although the major portion of the world's output of copper is utilized by the electric industry, copper also finds large use in the manufacture of alloys (brass, bronze, duraluminum) and antiparasitary products. Aluminum, which is the most abundant element (8.13%) in the earth's crust after oxygen (46.8%) and silicon (27.7%), is extensively used in the fabrication of the most disparate objects. Although not as good a conductor as copper, aluminum also finds large use in the electric industry because of its light weight and resistance to corrosion.[2] Tin is the basic material for the production of tinplate and several alloys of large industrial use. Nickel, platinum, and palladium are essential components of special materials and alloys. They are also widely used in catalytic reactions. Another metal that finds large use in the most disparate human activities is lead. Pipes, sheets, pigments, gasoline additives, products for agriculture, and drugs are some of the most important uses of this metal, whose global output is reckoned in many millions of tons per year. Just as important is zinc, which has similar applications except for gasoline additives, A by-product of zinc production is cadmium, which is extensively used in galvanization processes, in the manufacture of stabilizers for plastic materials, and so on.

*Depletion of Mineral Resources*

The consumption of metals is increasing at an exponential rate at the expense of limited resources.[3] It has been estimated that in 1770 world population was about 700 millions and metal consumption was 10% of that of 1900, by which time the world population had doubled. From 1900 to 1970 consumption increased 12-fold and population 3-fold. Clearly, we cannot go on at this pace. Reliable data show that deposits are being depleted and that industry is forced to work increasingly poorer minerals. This in turn entails greater difficulties and higher costs.

The industrial nations produce 66% of the world's minerals, but consume 90% of them.[4] The rate of production in the industrial nations is fast declining, however, and it is to be hoped that the developing countries will increase their production and exports of minerals. But in this field, too, as it has already happened for crude oil, a change in price policy can certainly be expected. One example will suffice to support this ominous prediction. In 1955 India imported about a ton of steel in exchange for each 15 tons of iron ore. At present it has to furnish 35 tons of high-grade mineral for each ton of steel. Such a situation, which depresses the economic growth of the developing countries, is bound to change.

Apart from cost considerations, it is the widening gap between resources and consumption that is so alarming. At the current rates of population growth and consumption, the world's mineral resources will be exhausted within 100–150 years.[4] Soon we shall have to turn to the problematic exploitation of undersea deposits,[5] the costly refining of low-grade minerals, and a more efficient recycling of metal products. (Of these, the last is already making a contribution to our supplies of certain metals, but awaits additional technological development to become significant.) With regard to the extraction of metals, there are some promising developments in the field of hydrometallurgy. Advanced research into products capable of fixing selectively particular metal ions, thus extracting them from ion mixtures, is under way in many laboratories. The products under study are generally polymeric compounds containing chemical groups capable of "chelating," or binding selectively, metal ions in solution. Since polymeric compounds are insoluble, the metal can be extracted from the solution. Preliminary experiments show that one day it may be possible to extract uranium ions from seawater or from the washing waters of phosphorites in which such ions are found in relatively low concentrations.

## Health Hazards

Among the various processes for the production of basic inorganic compounds, those that give the greatest cause for concern from the ecological point of view are the manufacture of sulfuric acid and chlorine-caustic soda. During the production of sulfuric acid not all of the sulfur dioxide ($SO_2$) produced is converted to acid; a fraction escapes into the air. Until the end of the 1950s the conversion was approximately 97% efficient, which meant that the gases emitted contained up to 3,000 ppm (parts per million) of sulfur dioxide. As a result there were frequent occurrences of "acid rain" across national boundaries. (Precipitations on Sweden, for example, contain sulfuric acid originating from German plants. This type of pollution is the frequent cause of international disputes.[6]) Legislation on atmospheric pollution has reduced the acceptable level of $SO_2$ in plant emissions to 500–1,000 ppm. This measure has caused some technical difficulties for the manufacturers as well as cost increases, but has brought about a substantial improvement in our environment.

Chlorine-caustic soda plants have been responsible for one of the best-known disasters caused by chemical contamination, namely, the collective mercury poisoning at Minamata, Japan. The electrolytic process commonly used in these plants entails the use of mercury cathodes. Part of the mercury is discharged into effluents and part in exhaust fumes. Mercury losses in chlorine-caustic soda plants are estimated at 1 ounce per ton of chlorine in water and half an ounce per ton of chlorine in the air. Before the world developed an environmental conscience or, more precisely, before scores of Japanese fishermen and their families were poisoned by contaminated fish from Minamata Bay, effluents were discharged into rivers and lakes without undergoing any purifying treatment. National and international health agencies have now taken steps to limit mercury emissions as much as possible, and as a result chlorine-caustic soda plants are undergoing drastic changes. In some cases, however, the changes required to implement current legislation are so extensive as to discourage the continuation of industrial activity in this field.

Iron and steel metallurgy involves processes that are fairly complex and involve very high temperatures, gas emissions, acids, and molten masses. Consequently, iron and steel works pose significant hazards, ranging from on-the-job injuries to exposure to toxic substances and the spread of acids and gases in the environment. With regard to the latter, there is a problem with the disposal of acids, and specifically of hydrochloric and sulfuric acids, both of which are used in production and, once used, must be either recovered or dispersed into the environment. Recovery, at least partial, is possible for hydrochloric acid, but not for sulfuric acid; the latter's disposal is therefore a very serious problem. An additional health hazard stems from the emission of gases that contain carbon monoxide and sulfur dioxide in addition to the more harmless carbon dioxide. Efforts are under way to reorganize this branch of industry in accordance with antipollution and job-safety regulations.

The production and spread of nonferrous metals also affect the environment and man's health. All ecological systems are contaminated in some measure by metals, which rank among the worst pollutants because they are wholly nondegradable. Few of them, moreover, are nontoxic, and even these can still interfere with the metabolism of other minerals biologically more significant. There is general agreement among scientists that many trace elements are essential to life; these metals are ingested in variable amounts and are often accompanied by nonessential metals, toxic and nontoxic. Table 6.2 lists the most common elements roughly divided according to their degrees of toxicity.[7]

Metals enter our environment partly from natural sources, partly from human activities. They diffuse into the air as particles and accumulate in waters and living organisms. With regard to particles, according to recent estimates[8] approximately 80% of atmospheric pollution from metals may be due to natural causes (dusts and forest fires) and 20% to human activities. Emissions from coal-burning plants are one of the major contributing factors (see chapter 7.)

Airborne particles enter the organism through the respiratory tract, while some liquids and gases may be absorbed through the skin. Particles larger than 1 micrometer [approximately 0.0004 inches] are stopped by the natural filters of the respiratory tract, but finer particles enter the lungs and cause considerable problems. Such particles are a vehicle for air pollutants and at the same time contain significant amounts of toxic metals that are made soluble by biochemical processes and then enter into the bloodstream. The consequences of inhaling such particles range from temporary irritations to lasting organic damage as in the case of silicosis, emphysema, chronic bronchitis, lung cancer, and other respiratory ailments. Metals can also be

**Table 6.2**
Toxicity of the most common elements

| Harmless elements | Toxic elements: relatively accessible | Toxic elements: insoluble or rare |
|---|---|---|
| Sodium | Beryllium | Titanium |
| Potassium | Cobalt | Hafnium |
| Manganese | Nickel | Zirconium |
| Calcium | Copper | Tungsten |
| Hydrogen | Zinc | Niobium |
| Oxygen | Tin | Tantalum |
| Nitrogen | Arsenic | Rhenium |
| Carbon | Selenium | Gallium |
| Phosphorous | Tellurium | Lanthanum |
| Iron | Palladium | Osmium |
| Sulfur | Silver | Rhodium |
| Chlorine | Cadmium | Indium |
| Bromine | Platinum | Ruthenium |
| Fluorine | Gold | Barium |
| Lithium | Mercury | |
| Rubidium | Thallium | |
| Strontium | Lead | |
| Aluminum | Antimony | |
| Silicon | Bismuth | |

absorbed by the organism through the digestive tract. Metals accumulate along the food chain and may reach dangerously high levels in the food we eat. Furthermore, foods can be contaminated directly through contact with compounds containing toxic metals.

Recent studies[9] have shown that several microorganisms found in the environment and at the bottom of the sea can methylate ions of toxic heavy metals such as mercury, palladium, thallium, lead, platinum, gold, tin, chromium, (as well as selenium, arsenic, and sulfur). The methyl compounds thus formed are soluble in fats and can enter the environment as well as the food chain. Biomethylation is a very important factor in the biogeochemical cycle of elements like tin, mercury, and arsenic since some bacteria can act even on minerals, that is, on chemical aggregates in which a metal ion is found in a form that makes it almost completely insoluble. As has been remarked, "Microorganisms synthesize these compounds more easily and efficiently than our best industrial chemists."[7]

While natural dusts are largely responsible for the spread of metals in particulate form, biomethylation contributes to the diffusion of heavy metals in the living world. This process is even more efficient when the ions are in solution, as in the case of effluents from chemical plants. A case in point, amply debated even in the popular press, is the mercury contamination in Minamata Bay. Briefly, this is what happened. In the anaerobic environment of the sea bottom, microorganisms methylated the mercury discharged by a chemical plant, converting it to monomethyl mercury chloride and dimethyl mercury. Absorbed by the microorganisms themselves, these compounds climbed the food chain through plankton and eventually accumulated in the fatty tissue of fish. Owing to the neurotoxic properties of organic derivatives of mercury (and many other metals), the prolonged ingestion of contaminated fish caused severe brain and nerve damage as well as many deaths among the inhabitants of the seaside village.[10] Worth discussing at greater length is a less known aspect of the Minamata tragedy, namely, the great difficulty and delay in ascertaining its causes.

## The Minamata Tragedy

*In April 1956 a girl 6 years old suffering from an unknown brain disorder presented a Minamata doctor with the problem of finding the cause of the ailment, as well as a cure. In the following month eight more people showed similar symptoms. A brain disease of unknown origin appeared to be spreading like an epidemic. In August, by request of the district governor, the School of Medicine of the University of Kammamoto organized a Minamata-disease study group, which in time would be augmented by experts from all over Japan. A first report released the following November suggested that the disease might be due to the presence of a heavy metal in the local seafood. Measures to limit its consumption were called for and great efforts were expended to analyze sediments and effluents from a nearby plant, but to no avail.*

*In 1959 there was a new and stronger wave of poisonings. Fishing in the district was forbidden and the Ministry of Health issued a preliminary report that concluded that the cause of the poisonings was an as yet unspecified form of organic mercury. Although the source of the compound was still unclear, there was a strong suspicion that the effluents from the chemical plant had something to do with it. The plant's managers, on the other hand, could prove that they were only using inorganic mercury, whereas the poisonings were caused by organic mercury. The solution of the enigma had to wait a few years. From 1960 to 1964 researchers at the University of Kammamoto succeeded in isolating methyl mercury derivatives (methyl mercury chloride and sulfide) in seafood and microbiological sediments taken from the bay. It could now be established that the inorganic mercury from the plant was methylated by marine microorganisms and then introduced into the food chain. In 1960 a purifying plant was installed at the factory and no more incidents have since occurred at Minamata.*

*No official steps were taken to control mercury contamination, however. As a result, another series of poisonings altogether similar to those at Minamata occurred in 1965 in the basin of the Agano river, downstream from a plant where acetaldehyde was produced by a process that requires the addition of water to acetylene in the presence of mercury salts. Subsequently, the Ministry of Health conducted a thorough inquiry that positively identified monomethyl mercury chloride as the cause of the collective poisonings and chlorine-caustic soda and acetaldehyde plants as the sources of mercury contamination. Following this inquiry, the government finally took action to prevent a repetition of such incidents.*

*It took 5 years to understand the problem, and another 5 to take the appropriate measures, which, however, did not put an end to the tragedy. Symptoms of mercury poisoning can appear years after exposure, and even today the number of casualties is not known with any precision. The official count is 234 dead and 1,300 poisoned, but according to Japanese experts the victims at Minamata and on the nearby islands number at least 10,000.[11]*

*People interested in the judicial process of other nations may be interested to know that not until March 1978 did legal proceedings start against Japan's prime minister and other ministers of the time, accused by a group of victims of not having taken steps to control and prevent water contamination by the Minamata chemical plant.*

The tragic story of the Minamata fishermen had such profound, worldwide repercussions at all levels of public opinion that all the international agencies have committed themselves to studying mercury from all possible environmental aspects. There are other metals, however, that have similar, if not worse, toxic properties and whose spread in the environment gives just as much cause for concern. Lead has been notorious for its toxic effects for a very long time. Saturnism (lead poisoning) was known in Roman times, and some scholars have even theorized that the decline of the Roman empire may have been accelerated by collective poisoning from the lead used in a great variety of products, such as cosmetics and water pipes.[12] Lead enters our environment by natural causes (dissolution of minerals, dusts, biomethylation, and so forth) as well as by human activities. The single greatest cause of the dispersion of lead is the use of one of its derivatives, tetraethyl lead, as an antiknock gasoline additive. Exhaust fumes from automobiles are veritable aerosols containing lead as well as carbon monoxide. Less harmful—although sometimes quantitatively more significant—sources of lead are batteries, electric cables, lead-based paints and ceramic glazes, and printed paper.

As a consequence of using leaded gasoline, pastures and cultivated fields adjacent to heavily traveled highways are contaminated by lead, as well as by other metals present in smaller measure in gasoline and tires. Table 6.3, which summarizes the results of studies conducted in the German Federal Republic,[13] gives an idea of the degree of this contamination. According to rough estimates, the amount of lead that is dispersed in Italy every year exceeds 70,000 tons. On the average, every Italian contributes 10 ounces (275 grams) to lead pollution through gasoline alone. On the basis of reliable calculations, it appears that the use of leaded gasoline has introduced enough lead into the atmosphere to cover every square yard of the Northern Hemisphere with about 0.3 ounces (10 grams) of lead.[14]

Beside contaminating the air, this toxic element is found in varying amounts also in water and in everyday foods. Analyses of the drinking water of 16 American states have revealed average lead levels of 0.00000035 ounces per quart (10 micrograms per liter). With regard to food contamination, the amount of lead found in the diet of the average New Yorker is illustrated in table 6.4. It should also be said that while almost all of the inhaled lead reaches the bloodstream, only 10% of ingested lead is absorbed by the body. Consequently, efforts

**Table 6.3**
Lead content of agricultural products grown along highways

| Product | Lead (ppm) | Distance (yards) from highway |
|---|---|---|
| Corn | 9.0 | 10 |
| Corn | 3.0 | 200 |
| Green lettuce | 16.5 | 10 |
| Green lettuce | 4.5 | 50 |
| Onions | 8.0 | 10 |
| Onions | 4.0 | 50 |
| Cauliflower | 8.0 | 10 |
| Cauliflower | 4.0 | 50 |
| Spinach | 11.5 | 8 |
| Celery | 4.2 | 8 |
| Carrots | 6.0 | 8 |
| Green lettuce | 44.2 | 2 |
| Parsley | 17.8 | 7 |
| Apricot (skin) | 7.5 | 5 |
| Apricot (meat) | 2.1 | 5 |

**Table 6.4**
Stable lead in New York City diet (1966 sampling)

| Diet category | kg/year[a] | mg Pb/kg | mg Pb/year |
|---|---|---|---|
| Dairy products | 200 | 0.04 | 8 |
| Fresh vegetables | 48 | 0.12 | 6 |
| Canned vegetables | 22 | 0.44 | 10 |
| Root vegetables | 10 | 0.07 | 1 |
| Potatoes | 38 | 0.17 | 6 |
| Dried beans | 3 | 0.02 | |
| Fresh fruit | 59 | 0.07 | 4 |
| Canned fruit | 11 | 0.25 | 3 |
| Fruit juices | 28 | 0.09 | 3 |
| Bakery products | 44 | 0.39 | 17 |
| Flour | 34 | 0.04 | 1 |
| Whole grain products | 11 | 0.13 | 1 |
| Macaroni | 3 | 0.08 | |
| Rice | 3 | 0.04 | |
| Meat | 79 | 0.42 | 33 |
| Poultry | 20 | 0.30 | 6 |
| Eggs | 15 | 0.22 | 3 |
| Fresh fish | 8 | 0.16 | 1 |
| Shellfish | 1 | 0.31 | |
| Total annual intake | | | 103 |

Source: John H. Harley, "Sources of lead in perennial ryegrass and radishes," *Environmental Science Technology* 4(3):225 (1970).
a. Abbreviations: kg, kilograms; mg, milligrams; Pb, lead.

are under way to reduce the permissible levels of lead in water, air, exhaust fumes, lead factories, and related industries.

Although most people are unaware of it, lead poisoning is a danger that touches everyone. To give an example, newsprint and certain inks and colors widely used in printing newspapers and children's books contain lead. Since children are apt to eat paper, especially if colored, this fact is a source of great concern to US health agencies. Epidemiological studies show a fairly high incidence of lead poisoning among American children. Obviously, this problem cannot be peculiar to the United States alone. The truth is that in other countries it is not properly recognized.

Another metal that has been under investigation for a long time is cadmium. In this case, too, as for mercury, legislative action has been prompted by an incident that occurred in Japan.

Itai-Itai Disease

*In 1946 Nobum Hagino of the Toyama district first became aware of patients with extremely fragile bones and suffering from excruciating pains in the joints. The unknown ailment was called itai-itai disease, or ouch-ouch disease, because of the screams of pain uttered by its victims. A brief epidemiological investigation narrowed the spread of the disease to the basin of the Jinzu River. Hagino's findings were reported at a meeting in 1955 (9 years after the first observations) and prompted academic studies on the disease that at the time was attributed to dietary deficiencies, in particular to lack of vitamin D, which is often suspected in cases of bone disease. Studies were performed on sediments and water downstream from a zinc mine located on the upper part of the Jinzu River, on rice fields, as well as on the autopsy reports of dead patients. It was suggested that the disease might be due to cadmium poisoning.*

*In 1961 (15 years after the disease had first come to Hagino's attention) the government of Toyama district and Japan's Minister of Health initiated a coordinated research program in which universities, local agencies, and hospitals were asked to participate. The project produced a voluminous amount of data, often contradictory. (Dr. Hirota from the hospital of the Kamioka Mines reported that animal tests with effluents from the mine had produced no evidence of damage.) Meanwhile, a group of victims and parents of dead patients sued Kamioka Mines for damages.*

*In 1968 the Ministry of Health issued a preliminary report that stated that itai-itai disease was to be ascribed to chronic poisoning from cadmium ingested through water and the rice grown in the Ginzu basin. In addition to the amount of cadmium normally found in rivers—particularly if they flow through areas where the soil is rich in cadmium—the water of the Ginzu was found to contain, in much larger measure, the cadmium discharged by the zinc mine. A certain amount of cadmium always accompanies zinc. Even though the amounts ingested were in themselves quite small, the prolonged ingestion of cadmium had caused a chronic intoxication whose symptoms appeared long after it had begun.*

*In 1969 (23 years after the first observations) the Japanese government issued regulations on the allowable levels of cadmium in water and foods.*

The question of the toxicity of nonferrous metals does not lend itself to definite conclusions. Many of these metals are toxic if absorbed in large amounts, but essential to life in trace quantities. Furthermore, the biological effects of one metal may be influenced by the concurrent absorption of another. Consider arsenic, for example. It is an element whose fame as a toxic substance has obscured its virtues as a growth agent;[15] in fact, it is commonly used as a feed additive. The case of selenium is even more interesting. While toxic in "significant" amounts, if absorbed in "low" amounts it seems to have the power to protect the organism against the action of carcinogens and also to act as an antidote to mercury. As in many other cases, the problem here is to define what is meant by "significant" and "low" with regard to the amounts of a particular metal absorbed by the organism.

Metals are everywhere, and their dissemination is due to natural causes as well as to industrial activities. The river Paglia, which runs past the mercury-rich rocks of Mount Amiata in Italy, contains significant amounts of mercury. Legumes grown in selenium-rich areas may contain as much as 4,000 ppm (parts per million) of this metal. The FDA has conducted a study on the institutional diet of 28 American colleges. Results show that it contains 0.003–0.01 ounces per pound (0.2–0.69 milligrams per kilogram) of antimony, 0.0003–0.001 ounces per pound of cadmium, and significant amounts of all metals considered toxic.

It is important to note that the most dramatic cases of metal poisoning—namely, mercury and cadmium poisonings—have been due to the monotonous diet typical of the poorer areas of Japan. A varied diet is therefore an important precaution for avoiding the prolonged absorption of metals, or any other toxic ingredient, contained in our foods.

# 7

## Energy and the Chemical Industry

Energy is, of course, essential to modern society.[1] The first benefit that man expects from the use of any form of energy is a measure of relief from manual labor: It has been estimated that 1 million Btu (British thermal units, equivalent to 252 calories) costs $6,000 if this amount of energy is produced by human labor (at $3 per work hour) and $0.25 if produced by fossil fuels.[2] The second is the possibility of converting more and more materials to useful products. Unfortunately, during the 20 years from 1950 to 1970 oil was so plentiful and cheap that "vast segments of society considered it as available as tap water."[3] This state of affairs has caused modern society to make the twofold mistake of savagely wasting oil resources and failing to encourage the development of alternative energy sources. What has been created, in effect, is an oil monoculture, by which I mean that both research and industrial production have been concerned with, and relied on, oil to the exclusion of almost everything else.[a]

### The Cost of Energy

Five industries absorb approximately 20% of the entire energy budget of a technologically advanced nation like the United States: (1) production of food; (2) conversion of raw materials—metals, glass, concrete, lime, and so forth; (3) exploitation of energy—in particular, oil and coal; (4) processing of chemicals (including plastics); and (5) papermaking.

#### Production of Food
Human labor and solar energy are not enough to produce food. Other forms of energy are needed, at least in our society. With reference to our nutritional system, and taking into account the energy needed to transport, preserve, stabilize, cook, and bring foods to our

tables, it can be shown that 7–8 calories of fuel energy are expended for each food calorie we eat. To put it another way, feeding technologically advanced people requires an annual consumption of 1,800 pounds of crude oil a head.

In the US model, agricultural production entails large use of fertilizers, pesticides, and machinery. High yields per acre and the resulting increase in production per farmer are obtained by expending energy in various ways.[4] In patriarchal systems (China, 1930), the food energy obtained is 46 times the energy expended; in the current preindustrial systems, it is 13–38 times greater; in the semiindustrialized systems of the tropics, it is only 5–10 times greater. In the modern industrial systems the input of energy is greater than the output. The ratio of energy output to energy input is 0.3 in Holland (1970) and 0.35 in England (1972). Turning now to specific types of production, wheat culture yields 3.4 calories for each calorie expended (England, 1970), corn culture 2.6 calories per calorie expended (United States, 1970), and potatoes 1.6 calories per calorie expended. The ratio is particularly unfavorable in the production of foods of animal origin: 0.37 for milk, 0.14 for eggs, and 0.1 for chicken meat (these data refer to England, 1970). The ratio is even worse for beef. Incidentally, the most expensive products in terms of energy are vegetables grown in winter in the northern countries in very special conditions. The ratio of energy output to energy input is 0.004 for winter tomatoes (Denmark) and 0.002 for lettuce (England). The production of proteins deserves special mention. A recent study[5] has given us information on the amount of energy needed to produce a gram of protein. In the case of seafood it is the energy expended for fishing. Herrings require 20,000 calories for each ounce of protein, wheat 41,000, rice 120,000, codfish 235,000, tuna 240,000, eggs 395,000, catfish grown in aquaculture 415,000, poultry 445,000, hogs 560,000, milk 790,000, and beef 2,400,000.

This energy is expended in various ways. Referring again to the US agricultural system, from which the European models are largely inspired, 18% of the energy is expended in the construction of buildings and services, 11% in machinery, 30% in fertilizers and insecticides, and 40% in fuel for trucks and electricity. The technological revolution in agricultural practices has had very positive results not only for total food production but for the individual farmer. A small number of farmers can feed a great many people by utilizing a limited expanse of land. Farm income has risen, even though the market prices of the leading commodities are essentially low. Furthermore, manual labor is only 5% of the energy input in modern agricultural systems, while it constitutes 95% of the energy expended in patriarchal systems and 40–60% in semiindustrial systems.[6]

*Conversion of Raw Materials*

Turning a raw material into a usable material entails processes of transformation and purification that require the expenditure of energy. Largely disregarded before the 1973 energy crisis, this factor has now become a primary consideration in every industrial activity, and quantitative studies on the energy implications of materials processing are becoming more numerous and accurate. Knowledge of these aspects is essential in order to modify the processes themselves in accordance with the new policy of energy conservation and to make a rational plan for the production of energy.

Metals are produced from minerals by processes that require large amounts of energy—amounts that vary depending on metal concentrations. According to a study by Battelle Columbus Laboratory, in 1973 about 8% of the entire energy budget of the United States was devoted to the manufacture of the ten most important metals, namely, iron (and steel), aluminum, zinc, lead, copper, chromium, manganese, magnesium, titanium, and cadmium. A ton of iron requires the expenditure of 30 million Btu, aluminum 240 million, zinc 65, copper about 100, magnesium 350, and uranium (as oxide) 800–1,000 million. These energy costs are bound to increase as the metal content of minerals decreases. At present many manufacturers are trying to relocate the more energy-intensive processes in the mineral-producing countries if, as is often the case, these countries are rich in energy as well.

Apart from metals, modern technology employs a number of inorganic materials such as lime, concrete, ceramic, glass, and refractories. These materials do not require extensive processing, and therefore their energy costs are relatively low and due essentially to mining, transportation operations, and, in some cases, calcination. However, since they are used in very large quantities, their contribution to the total consumption of energy is significant, about 2% of the total. A ton of lime costs in energy 8.5 million Btu, concrete 7.6, ceramic 3.5, glass 17, chalk about 1, and sand 0.05.

*Exploitation of Energy*

The extraction, conversion, and transportation of energy is a very costly process—indeed "the most energy-consuming sector of the American industrial scene. More than 20% of the gross energy of fuels is consumed in bringing energy to the materials processor."[3]

The United States relies heavily on coal for its energy needs. Since American coal fields are fairly rich, mining operations are relatively easy. Consequently, the energy expended to make coal available does not exceed 2–3% of the energy content of coal. Turning to oil, we find that the ratio is not so favorable: About 4.7% of the US energy budget is devoted to oil production and refining,[3] which is equivalent to saying that operations ranging from the exploration of oil fields to the distribution of refined products consume 17% of the energy content of oil itself.

The expenditure of energy associated with the production and distribution of electricity is also very substantial. Some *two thirds* of the gross energy of a fuel is consumed in producing and transporting electricity. A large part of the energy is lost as heat, which in turn causes problems of thermal pollution. Unaccountably, only a very small fraction of this heat is currently being utilized, mostly in the agricultural sector (for example, in greenhouses).

## Processing of Chemicals

Petroleum is the key ingredient in most of the products of industrial organic chemistry. Some of the most important petrochemicals are ethylene, acetylene, propylene, butadiene, benzene, and toluene. The production of these compounds and of the plastic materials made with them consumes as much as 2.5% of the entire US energy budget. The plastics industry uses more energy than even the very profligate aluminum industry. Part of this energy can be recovered by burning discarded plastics; however, it is only a small fraction, owing to the extreme dispersion of plastic products and the difficulty of collecting them, which effectively prevent the development of economical recycling techniques. Indeed, according to one study, out of 15 million tons of plastic materials produced in the United States in 1974, only 1 million tons were recovered and recycled.[7]

## Energy Production

One conclusion is immediate: The production of energy is the key problem for mankind. We must solve this problem in order to eat, manufacture products, travel, and heat ourselves. We must solve it while confronted with many other no less pressing problems—the pollution of our environment, the depletion of our mineral resources, and the progressive erosion of our soils. The solution of the energy problem is certainly hampered by the extreme rigidity of our political and economic systems. In a lucid essay Nathan Keyfitz remarks, "Social structures are as solid a reality as raw materials. After two years of discussions on the need to conserve their natural resources, the United States continues to consume the same amounts of fossil fuels. Confronted with a global shortage of raw materials every person of good will wants to see every form of waste reduced, but . . . the enormous viscosity inherent in the models of production and consumption makes the reduction of consumption an improbable instrument. . . ."[8] This conclusion raises a question that is neither pessimistic nor far-fetched: "Will there be enough hydrocarbons for agriculture?" If not, we "must face an eventual return to horse-and-buggy agriculture, with all that it entails in the form of a large movement of people back to the farm and to hard manual labor."[9]

At present the principal sources of energy are oil, coal, natural gas, and hydroelectric and nuclear power. Figure 7.1 illustrates world production of the primary energy sources and their respective rates of increase.

### Energy from Oil

In 1970 petroleum supplied more than half of the energy needed by most of the industrial nations. The major cause of this situation, which in effect created an almost total dependence of modern industry on oil resources, was the low cost of this form of energy from 1948 to 1973. A barrel of oil cost $2.6 in 1948 and $3.39 in 1972. If we take inflation into account, in 1972 a barrel of oil actually cost $1.85, that is, *less* than in 1948.

From the environmental point of view, all the operations involved in the extraction, transportation, and refining of oil cause fairly serious problems. Oil wells are susceptible to fires and losses. Offshore drilling is particularly hazardous because oil spills are extremely difficult to control. The transportation of oil causes the systematic pollution of the oceans. In addition to oil spills due to loading and unloading operations, the washing of tankers is a considerable source of pollution; this operation is usually done at sea during the return trip (after the tanker has unloaded its cargo), all international agreements and conventions notwithstanding. Mishaps at sea and shipwrecks are another major contributing factor.

Two independent studies conducted by the University of Oklahoma and the NAS (National Academy of Sciences)[10] estimate the direct pollution of the sea from hydrocarbons at 35 million barrels. In addition to direct pollution—due to operations at sea, losses during transportation, accidents, dumping from refineries, and so on—there is an indirect pollution caused by hydrocarbons that enter the atmos-

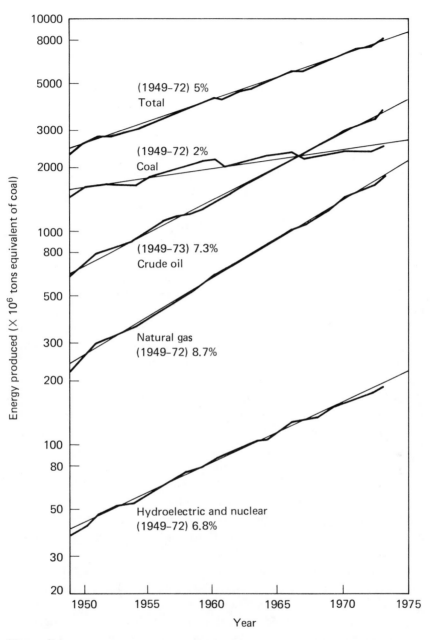

**Figure 7.1**
World production of the primary energy sources and their respective rates
of increase. (From *The World Symposium on Energy and Raw Materials, Paris,
1974*)

phere from various sources (such as evaporation and unburnt residues) and eventually end up in the sea. While the two studies agree in their estimates of direct contamination, they differ substantially in their evaluations of indirect contamination: the University of Oklahoma estimates it at 63 million barrels per year, the NAS at 432,000 barrels per year.

This type of pollution has various consequences, all decidedly negative. To begin with there are esthetic and economic considerations. Oil slicks stretching across the sea, beaches covered with tar, and coastlines darkened by naphtha are not only ugly but an economic threat to fishermen and resort areas. In addition, hydrocarbons cause more or less extensive changes in the marine ecosystem. By disrupting the normal development of plankton they cause a decline in marine fauna. Last, there is a very real danger to man's health from the contamination of seafoods. Combating this type of pollution is expensive. According to one study,[11] cleaning up a spill of 35,000 tons of oil (a not unusual amount in the event of a tanker's breakup) costs $6–8 million.

The combustion of oil constitutes an even greater hazard in that it affects man directly. Oil fires cause deaths, burns, and permanent and temporary disability. More important, burning oil and its derivatives is a major cause of air pollution. Power plants, heating furnaces, and gasoline engines spew into the atmosphere significant amounts of sulfur oxides, nitrogen oxides, carbon oxides, unburnt residues, and polynuclear aromatic hydrocarbons. Heavy metals (and especially lead in the case of gasoline) are also emitted in large quantities. To have an idea of the toxicity of this type of emission recall what happens when somebody, either intentionally or by mistake, leaves the car engine running in a closed garage. The most serious consequences of air pollution from the combustion of oil and the interaction of pollutants with ozone under the action of ultraviolet radiation (photochemical smog) are asthma and other respiratory ailments, the growing incidence of lung cancer, and chronic poisoning by heavy metals. It is not only man's health that is affected, however, but his property as well. Buildings are soiled, metals corroded, and historical monuments eaten away by sulfur dioxides and acid compounds originating from motor vehicles and home furnaces. Something has been and is being done at the legislative level. Antipollution laws have set restrictions on the sulfur content of heating oil, the amount of lead that may be added to gasoline, and the emission levels of carbon and nitrogen oxides from automobiles. Unfortunately, all such antipollution measures conflict with the need to save energy created by the current crisis; besides, they are quite expensive. As a result, deadlines for their implementation have often been allowed to lapse.

*Energy from Coal*

The king of fuels until World War II, coal has been supplanted by oil and natural gas for four main reasons: It is difficult to mine, dangerous to produce, expensive to transport, and awkward to use. At present, however, there is a general rethinking, particularly in those countries such as the United States, Germany, and England that have large supplies of coal. Coal reserves already discovered or accessible without excessive cost amount to 200 billion tons in the United States, 24 billions in the German Federal Republic, and 3.5 billions in England, for a total of 250 billion tons in the OECD countries. As for untapped reserves, there is general agreement on an estimate of approximately 670 billion tons, which is equivalent to saying that the United States has enough coal reserves to meet its energy needs for 113 years, provided they remain constant at the 1972 level. For the whole OECD area reserves are sufficient for 51 years.

All experts agree that the use of coal for electric power generation will increase dramatically in the years to come. However, before coal can become a major fuel again some formidable environmental problems must be solved. Mining costs in the United States have risen substantially since the passage in 1969 of a law on occupational safety and health in coal mines aimed at reducing accidents (often fatal) and job-related diseases as much as possible. As a result, coal operators in the United States and other coal producing countries are quickly switching from underground to surface mining. Strip mining causes severe environmental problems, however, both from the esthetic point of view and because of acid runoff, silting, and the disposal of debris. In addition, by destroying prime farm land, it often conflicts with agriculture.[12] Part of the damage can be corrected by restoring the land to its original conditions after mining operations have been completed, but land reclamation is expensive. Some figures are available: The cost of reclaiming a piece of land in the Rhineland after the exploitation of a lignite field was $3,000–4,000 per acre. The same amount was spent in England for a similar operation, while in the United States costs range from $100 to $4,000 per acre, depending on the type of works.

Generating energy from coal creates in turn a number of problems. Coal burning, which is an essential part of the power-generating process, produces substantial emissions of sulfur oxides, carbon oxides, and heavy metals. According to a study by the American Chemical

Society,[13] most of the volatized mercury enters the atmosphere. Data released by the EEC set the worldwide input of mercury into the air through coal burning at 3,000 tons per year, 90 of which come from the German Federal Republic and 65 from France. By a very rough estimate, mercury emissions in Italy would amount to 20 tons per year. With regard to cadmium, the same EEC study estimates emissions in the German Federal Republic and in France at 90 and 65 tons per year, respectively. The figures are based on the average amount—about 1 ppm (parts per million)—of cadmium and mercury found in coal and on volatization measurements.

Coal fly ash also poses a very serious problem. It contains selenium and arsenic and heavy metals such as lead, cadmium, nickel, and chromium and is emitted in very large amounts. "It is estimated that even with 99% efficiency fly ash removal, approximately 7,000 tons of ash are discharged into the air each year from a 3,000-MW [3,000-megawatt] coal burning station."[14] C. E. Chrisp and his coworkers[15] have found evidence that coal fly ash contains mutagenic substances. Consequently, the carcinogenicity of respirable fly ash is a question that must be carefully evaluated, particularly in view of the amounts involved. To quote the same author, "It can be estimated that in 1974 a total of $2.4 \times 10^6$ metric tons of fly ash was released to the atmosphere from all coal-burning facilities in the United States."

Complying with the new regulations on emission levels from coal-burning power plants entails extensive modifications of the facilities and considerable expense. As the time to implement antipollution regulations draws near, opposition from the electric utilities intensifies, and disagreements, discussions, and changes of mind are the order of the day. It has been proposed, for example, that coal-burning facilities be located in the populated areas where the energy will be used and, at the same time, that acceptable standards for power plants or for coal grade be lowered or implementation of the norms be postponed.

Owing to the complexity of the issue and the difficulty of establishing truly meaningful parameters, the differences of opinion have led to the creation of veritable battle camps, much to the detriment of fruitful study and discussion. A series of meetings between environmentalists, industrialists, and representatives of health agencies have been planned in the United States in an effort to reach an agreement on general standards for an intensive utilization of coal, a realistic definition of acceptable pollution levels, and precise operational norms. The opposing sides are so far apart that it has been felt necessary to invoke the "rule of reason" as a guiding light for the project and to set down a code of behavior for the participants. As L. J. Carter remarks, "It reads almost like a Boy Scout code: 'They [project participants] will share all pertinent facts; they will not mislead each other with unfair tricks; they will not lightly impugn each others' motives; they will avoid dogmatisms; . . . they will distinguish between facts and value judgments.'"[16]

## Energy from Natural Gas

In 1970 gas—essentially methane—supplied 34% of the energy needs of the United States, 10% of Italy's, and 7% for the whole of the OECD countries. With regard to the immediate and medium-term outlook for the supply of methane, the same considerations apply as for oil. Even though there are excellent prospects of finding new deposits, gas by itself cannot be the source that will enable us to meet our energy needs in the next 30 years. In a few countries, however, and particularly in the United States, methane will continue to be a considerable source of energy for at least 10 years.

The extraction, transportation, and use of natural gas entail operations that are not always free of risk for either man or the environment. Fires and explosions are an ever present danger while drilling or operating wells, as well as during transport, whether this is done by pipeline or tankers. Underground and aboveground losses during transport are relatively minor risks. The combustion of methane for electric power generation has the same drawbacks we have discussed for oil, although to a lesser degree.

## Hydroelectric Energy

This form of energy production is strictly tied to the availability of water resources. Generally speaking, hydroelectric energy constitutes less than 2–3% of the energy budget of industrialized nations. There are no drawbacks to this type of power generation that are in any way associated with chemistry, nor are there problems of environmental pollution. This is not to say that hydroelectric energy is devoid of drawbacks.

The exploitation of water resources for the production of energy entails the creation of reservoirs or artificial lakes and the construction of dams. Such modifications of the natural environment may cause ecological upsets. The construction of the Aswan dam, for example, in the opinion of various experts has resulted in a drastic decrease in marine fauna in the area of the Mediterranean around the estuary of the Nile because of the decline in the amount of lime that is carried out to sea. There is also a very real danger of serious

disasters associated with these works. According to one study, 33
dams collapsed in the United States in the period 1918–1958, for a
total of 1,680 dead.[17] It has been reported that from 1959 to 1965
9 other dams collapsed (in addition to a larger number of minor accidents).[18] In 1976 6 accidents occurred for a total of more than 700
dead. In Europe we recall the Frejus disaster, in which hundreds of
people were killed, and the 1963 Vajont disaster, which claimed the
lives of 2,000 people. In the latter accident the dam withstood the fall
into the reservoir of a huge landslide. Construction defects, faulty
geological surveys, storms, landslides, and earthquakes are the most
common causes of such disasters, which make hydroelectric power
one of the costliest energy sources per kilowatt-hour produced in
terms of human lives and property losses.

*Nuclear Energy*

In 1970 nuclear power contributed less than 1% to the energy needs
of the industrial nations. Yet a great deal of the scientific community
agrees that this form of energy is the only one that can replace fossil
fuels, despite its many technical difficulties and high investment costs.
The slow development of nuclear power in recent years is due mainly
to two facts. The first is the shortsightedness of our political and technological leaders who have placed all their faith in an unlimited supply of cheap oil. Italian plans for nuclear power plants, for example,
were abruptly shelved by a hostile political world convinced of the
lasting technical and economic superiority of oil energy.[b] The second
obstacle is the strong opposition it encounters in many quarters because of environmental considerations.

The construction and use of a reactor, the production of nuclear
fuel, and the disposal of waste material pose very serious safety problems. During the production of nuclear fuel there is a danger that
radioactive dust may be released into the environment and that
radioactive waste may contaminate soils and groundwaters for long
periods of time. Shipping nuclear fuel is also hazardous. Furthermore, there is the danger of accidental losses or thefts, which can only
be prevented by strict security measures enforced by highly
specialized and costly personnel. Operating a reactor requires in
turn numerous safety measures. Some of these problems have not
been satisfactorily solved as yet, notably, the question of radioactive
waste.

A further cause for concern, of a political nature, is the relation
between the peaceful use of nuclear power and the proliferation of
nuclear weapons. We know that the plutonium produced as a waste

product in reactors can be purified and used to make atomic weapons. It is feared, therefore, that an increase in nuclear power plants may lead to an increase in the production of weapons. It can be pointed out, on the other hand, that nothing prevents the military from manufacturing their own plutonium in ad hoc reactors, just as nothing thus far has prevented many nations from joining the nuclear-weapons club in spite of the various treaties of nuclear nonproliferation.[19]

If widespread concern over the safety of nuclear power has greatly slowed down the construction of nuclear power plants, it has also caused existing facilities to be built with the greatest precautions. As Manson Benedict of the Massachusetts Institute of Technology (MIT) states, "No member of the US public has ever been injured, much less killed, as a result of accidents in nuclear power plants. . . . Nevertheless, it is recognized that there is a finite probability of accidents in nuclear plants."[20]

The chance of accidents appears to be relatively small. According to an MIT study, the total effect of the entire spectrum of possible accidents might lead to an average of 0.8 accidental deaths and 8 injuries per year for the production of 200 billion kilowatt-hours of nuclear electricity. A study by the American Physical Society would increase these numbers by a factor of three, mostly because of the long-term effects of exposure to radiation. Even so, the effects are small compared with the effects of generating energy from coal, which recent studies estimate at 6,000 deaths per year for the same electric output. By taking all industrial and civilian consequences into account, a different MIT study arrives at a figure of 11 deaths per year for the same electric output. D. J. Rose comes to analogous conclusions: "What seems increasingly clear is that the hazards of burning fossil fuels are substantially higher than those of burning nuclear ones, yet many debates have enticed the uncritical spectator to just the opposite conclusion."[21]

It should be pointed out, however, that the above figures are the result of probabilistic computations and that the reliability of all such projections is a matter of opinion. It is undeniable that a nuclear accident might have very serious consequences. On the other hand, we have learned from experience that an accident involving a dam can also be very serious.

With regard to our ability to supply reactors with sufficient nuclear fuel, a report of Italy's National Research Council (CNR) on the country's energy needs states that "current supplies of the primary nuclear source are sufficient to feed all the thermal reactors that the power-plant industry can build in the coming years until the advent of fast breeders. With the latter the cost of the primary source will be so low . . . as to consent an almost unlimited expansion in energy production."

Research in the field of reactors' safety, fast breeders, and, long term, of controlled nuclear fusion is one of the few hopes mankind has of solving its energy problems.[c]

### Energy from Developing Technologies

Excepting nuclear power, the technologies for power generation discussed so far have been known and tested for a long time (though for these, too, better-documented criticism may lead to legitimate risk concerns). In addition to these technologies, others currently under research and development seek to exploit geothermal and solar energy sources and also the energy associated with winds and ocean thermal gradients. Impetus for this has come from the energy crisis, which forced the international scientific community to come to grips with the need to develop alternative energy sources.[22] (One of the most serious consequences of the energy "monoculture" created by plentiful supplies of cheap oil and natural gas has been the slowdown in the research and development of energy sources that would decrease in significant measure our dependence on fossil fuels.) In our current situation, which requires research efforts capable of quick results, geothermal and solar energy are expected to make a significant contribution to our energy needs, while the other technologies should not be given much relevance in any emergency planning, according to the energy report of the Italian National Research Council.

Geothermal energy is the form of energy that can be obtained by tapping the enormous reservoir of heat located at varying depths under the earth's crust. Geothermal power plants are already in existence, but exploitation of this form of energy up to now has been limited and quantitatively insignificant (total producing capacity does not exceed 1 million kilowatts). Three major geothermal facilities are at Larderello (Italy), Matsukawa (Japan), and the Geysers fields near San Francisco.

While there is general agreement that the amount of energy stored inside the earth is immense, opinions vary widely as to the fraction of this energy that can be tapped and when. Research is currently under way in the United States and elsewhere into the possibility of exploiting deposits of heated rocks deep underground. This can be done by drilling a well in which water is pumped down to the layer of heated rocks, thereby converting the water to steam, which is then brought to the surface through a second well. Although this technique is very promising, there are still some unanswered questions, such as drilling costs and the chance of triggering earthquakes, as well as some environmental problems due to the large amount of water and debris involved in such an operation. Generally speaking, geothermal wells have some potentially harmful environmental effects, such as water contamination from mineral wastes, seismic phenomena, and possible emission of hydrogen sulfide.

It has been calculated that the amount of solar energy that reaches the atmosphere is 1 kilowatt-hour a minute per square yard. Although part of this energy is absorbed and dispersed, a good half reaches the surface of the earth. Harnessing the sun's power is one of the challenges confronting modern technology. At present solar energy is utilized by man in two ways: indirectly, through photosynthesis and the formation of biomasses; and directly, by means of physical systems.

Photosynthesis is the process by which plants convert solar energy, carbon dioxide and water to biomass. By comparing the amount of solar energy that reaches a plant with the amount stored in it as chemical energy, it can be shown that in nature solar energy is converted to chemical energy with an average efficiency of 0.1–2%. Some plants, notably sorghum, *Hevea,* and sugarcane, are more efficient than others at storing energy. It is estimated that *Hevea* can produce annually a ton of rubber per acre, while sugarcane furnishes on the average 4 tons of sugar, which is equivalent to 2 tons of ethyl alcohol and 1.2 tons of ethylene. Many large research groups are currently studying ways of using the energy stored in biomasses by direct burning and by producing biofuels such as methanol through microbiological processes. Beside being a fuel, methanol can be a precious raw material for the production of ethylene. Because of the increase in the price of oil and the resulting increase in the price of petrochemicals, the production of ethylene from biomass has now become economically feasible.

Solar energy can also be tapped directly, by means of devices designed to collect it and convert it to heat. Solar heating systems have already achieved wide currency in technologically advanced countries like Japan and Israel, but ways of using this form of energy are currently under study in many other nations. Solar pumps, collectors, and batteries are some of the devices that have proved quite efficient. Practically speaking, however, no technology for the economical conversion of solar energy to electricity exists as yet.

While prospects for using solar energy stored in biomasses are optimistic, solar devices are not expected in the near future to supply more than a small percentage of our energy needs. On the other hand, these energy sources have very slight ecological and environmental drawbacks. A sound management of the plantations and a few precautions during combustion (mainly to prevent emission of fumes) would be sufficient to eliminate the problems.

# 8

## The Babel of Tongues

Several decades ago Paul Valéry wrote; "Never has humanity known so much power and so much confusion, so much worry and so much play, so much knowledge and so much uncertainty. In equal measure, does now anguish, now futility, command the hours of our days."[1] These words apply in an uncanny way to the current situation. In the two decades from 1950 to 1970 Science knew Certainty, if not in reality, at least in the popular view. In those 20 years the world viewed Science as an all-powerful entity, a sort of mother goddess of modern times. From it, people expected drugs for sleeping and for forgetting their troubles, synthetic milk and automobiles, insecticides and dishwashers, weekends on the Riviera or on the moon. Science fiction painted an automated, perfect world inhabited by beautiful, elegant, and noble people. If ever there was a problem in such a world, it usually came from a discovery that could not be controlled, a discovery so perfect that it exceeded the expectations of its creators. Never was there a hint of a future shorn of resources, of a world in difficulty, of a new Dark Age.[2]

A technologically advanced society had conquered Space and walked on the moon—objectively, an extraordinary achievement. It was reasonable to expect everything from Science, from Technology, from Them, from all these capital letters. Everything was going to be beautiful, just, organized, coordinated, and managed, with no unpleasant surprises. If, more realistically, we now turn from capital to lowercase letters, we realize that They, Science, and Technology do not exist. What does exist are politicians, scientists, technicians and experts, each with his own culture, ability, shortcomings, altruism, ambition, political concerns, shortsightedness, wisdom, and petty compromises. And what we see are unbelievable blunders. Take the energy crisis, for example. In retrospect, it was predictable, inevitable. The same for the environmental crisis. In a world of finite resources and frenzied technological progress the economic crisis of a system based on an annual, *compounded*, growth rate of 5–7% was as inevitable as night follows day.

For years, for decades, we proceeded as if nothing was going to stop us. When the crisis came, some were rewarded and others suffered. Since 1973 the large oil companies have seen their profits rise astronomically, while the Third World and the Fourth World have not been able to compete with the affluent nations for the purchase of needed grains and fertilizers. With regard to the production and sale of chemicals we have acted in a manner that is certainly open to criticism. Our basic policies have been generally at fault. The problem of the utilization of natural resources has not been properly addressed, and as a result an irreplaceable heritage has been squandered. For the sake of immediate advantages the long-term toxic effects of chemicals have not been sufficiently considered. Benefits have been overstated, risks ignored. We have created the expectation of a world without hunger, disease, want or pests.

Suddenly, the rude awakening, and we go from one extreme to the other. While only a few years ago we accepted everything— everything was good; the synthetic (made familiar, though never made comprehensible, by mysterious symbols like DDT, PCB, and BHA) was better than the natural—now we reject everything; we want to return to the natural; we rhapsodize about the "good old days," of which we forget the worst aspects.

The good old days could deservedly be the subject of a critical examination, but in this book I shall confine myself to a few voices speaking to us from the past on the questions of insects. Herodotus thus described the nights of the Egyptians in book II of his *History of the Persian War*:

Against the insects that afflict Egypt in countless swarms they defend themselves with various expedients. People who live north of the swamps climb on top of their towers where they can sleep unmolested because mosquitoes cannot fly so high on account of the winds. The inhabitants of the swamps have devised other ways. They all own nets that they use for fishing during the day and at night they drape over their beds; then they crawl under them and thus protected they sleep. Wrapping oneself in a cloak or blanket does not help because mosquitoes can sting even through fabrics, whereas they do not even try to slip through the holes of the net.

In another part of the world people seem to have fared even worse. In Venezuela, we are told, "At the time of the conquest by Europeans, the highland tribes were superior in every way to those of the low-lands. Perhaps the cooler air of the mountains was more conducive to active life; but there were many other factors which may have been important; such as better water, better diet, . . . and a relative freedom from the mosquitoes and other insects of the lowlands."[3] In more re-cent times, we should remember that around the turn of the century 200,000 new cases of malaria were reported in Italy every year, "which suggests that at least a fifth of the Italian population may have been afflicted by this debilitating disease."[4] Obviously, if we do not know or forget all of this, the use of insecticides may appear less nec-essary or desirable in 1979 than it was at a time when flies infested our houses and mosquitoes spread malaria not only in the Third World but in the Pontine Marshes.

We have not been quite so successful in taking care of locusts. As the Bible tells us (Exodus 10:14),

And Moses stretched forth his rod over the land of Egypt, and the Lord brought an east wind over the land all that day and all that night; and when it was morning, the east wind brought the locusts. And the locusts went up over all the land of Egypt, and rested in all the coasts of Egypt: very grievous were they; before them there were no such locusts as they, neither after them shall be such.

For they covered the face of the whole earth, so that the land was darkened; and they did eat every herb of the land, and all the fruit of the trees which the hail had left; and there remained not a green thing in the trees, or in the herbs of the field, through all the land of Egypt.

A few thousands of years later locusts are still with us. Here is a de-scription of a recent invasion of locusts in the Horn of Africa:

This time the locusts have landed in Ethiopa, a land already troubled by 4 years of constant political strife, war, and guerrilla warfare. In the middle of June FAO agronomists counted 33 different swarms in the northern provinces of the country, each extending over an area of 10 to 180 square kilometers. In neighboring Somalia 17 swarms have already been sighted. Every swarm—millions upon millions of *Schis-tocerca gregaria* Forsk, as African locusts are called—moves about slowly, leaving a desert in its path. Grass, leaves, fruit, shrubs, and bushes—everything is destroyed and devoured by these terrible pests, which can strip 250 square kilometers of cultivated land in a single night.[5]

A few years ago public opinion was suddenly awakened to the dan-gers of chemicals by a series of serious accidents that were a sign of a

dangerous general situation. A number of measures were subsequently taken, but they were neither well thought out nor well
coordinated. Indeed, the problems initially were tackled in a somewhat haphazard manner, even within international organizations. It
must be pointed out that in the matter of hazardous chemicals public
health agencies have always taken action *after* the occurrence of a serious incident. Mercury contamination is a case in point. As mentioned
in chapter 6, more than 10 years went by from the first evidence of
intoxication at Minamata before the Japanese government took official steps to regulate mercury levels. Since then, all kinds of organizations, large and small, have concerned themselves with mercury,
among them the European Economic Community, the Organization
for Economic Cooperation and Development, the US Environmental
Protection Agency, and the World Health Organization.

After the cadmium disaster the same agencies threw themselves
into the study of cadmium, all at the same time and without any attempt at coordinating their efforts. The result, as could be expected,
is that we now know a great deal about some substances and very little
about others. Furthermore, it is not at all clear that priority was given
to the most serious problems. The only reason why mercury received
so much attention is that the Minamata tragedy had an enormous
inpact on public opinion. And even when measures were finally taken
to control mercury contamination, they were surprisingly incomplete.
While the strictest limitations have been imposed on industrial
effluents, not much attention, practically speaking, has been paid to
the large quantitites of mercury and cadmium that are released into
the air by coal-burning plants (see chapter 6). Yet mercury, although
toxic, is certainly less harmful than some other elements or compounds not yet thoroughly investigated.

Similar considerations apply to dioxin. The 1974 Seveso incident
brought dioxin to the attention of the whole world. Yet the potential
hazards of the production of trichlorophenol were already known in
1971. It took a tragedy of major proportions to call the authorities'
attention to a problem that is extremely serious from any point of
view. Preventive measures taken subsequently, moreover, were not as
thorough as they should have been. The consequence is that people
now see dioxin everywhere, even where there is none.

I cannot help observing with a certain dismay that while allowing
the use of highly toxic compounds containing lethal impurities, health
agencies have persecuted certain processes or products beyond anything that is reasonable. The ban of saccharin and cyclamates on evidence that many consider circumstantial may be justified in absolute

terms, but seems something of a mockery in light of the negligence that has prevailed, and still prevails, in many other areas. In addition, such clamorous episodes force many scientists into lengthy debates, thereby taking time and precious resources away from the consideration of problems that are objectively more serious.

In the Babel of tongues of this first phase, a pattern has been set: As a result of an accident, a press campaign, or a legal action, one of our countless problems suddenly becomes problem number one. Great research efforts are expended, production is halted, products are taken off the market; then everything stops short of the next step, namely, the establishment of rational standards and the study of truly vital concerns. In effect, such episodes are used as scapegoats for a bad conscience and often serve to appease public demand for environmental action. Hence incongruities are the order of the day: We have already mentioned some; I shall now present three more.

## Tobacco

Tobacco kills four times as many people as car accidents. This astonishing fact, established by Britain's Royal College of Physicians, was reported in the EEC publication *Euroforum*.[6] A survey in a British general hospital revealed that the percentage of "cigarette beds" (that is, beds occupied by people suffering from diseases due primarily to smoking) was about 10 per 100. According to *Euroforum*, "Between one third and a half of smokers will die from illnesses linked to smoking. . . . In France, tobacco kills about 35,000 people per year prematurely—15,000 of these through cancer. It reduces the life of a normal smoker by about nine years. The main causes of death in order of priority are: lung cancer, bronchitis and emphysema, and cardiovascular illnesses. Finally the use of tobacco during pregnancy can have disastrous consequences: twenty-percent of miscarriages can be attributed to smoking."[6]

The fact that tobacco is harmful has been known for a long time,[a] but it is only recently that the dangers of smoking have been put in the proper prospective. Its toxic effects are neither immediate nor spectacular, as in the case of, say, mushroom poisoning. It takes years for the toxic substances contained in cigarette smoke to cause chronic bronchitis or the development of lung cancer. Smokers have also been known to get sick years after they stopped smoking.[b]

The chemical analysis of cigarette smoke is sufficient to give a good idea of the hazards of smoking even to people who are not experts in toxicology; table 8.1 lists the principal components of cigarette smoke.

**Table 8.1**
Composition by weight of cigarette smoke

| Compounds | Percentage | Origin |
|---|---|---|
| Nitrogen | 59 | Air |
| Argon | 1 | |
| Oxygen | 13.4 | |
| Carbon dioxide | 13.6 | Products of pyrolysis |
| Carbon monoxide | 3.2 | in gaseous state |
| Water | 1.2 | |
| Hydrogen cyanide | 0.1 | |
| Hydrogen | 1 | |
| Various organic compounds (aldehydes, ketones, hydrocarbons) | 1.4 | |
| Water | 0.4 | Products of pyrolysis |
| Organic acids | 0.14 | in the aerosol state |
| Glycerol and alcohols | 0.1 | (condensed) |
| Aldehydes and ketones | 0.1 | |
| Hydrocarbons | 0.08 | |
| Phenols | 0.03 | |
| Nicotines | 0.04 | |

Source: P. Freour and P. Coudray, *Fumeurs en question,* Paris (1977).

Their effects on man's health are unquestionably bad. Nicotine is a highly toxic alkaloid that acts on the nervous system. Injecting a small amount of nicotine can kill a warm-blooded animal. While very potent intravenously, it is scarcely active orally. When smoke is inhaled, nicotine is drawn into the lungs, whence it passes into the capillaries and the bloodstream. Cardiac rhythm increases, blood pressure rises, and peripheral circulation decreases. The secretion of gastric juices and the proper functioning of the stomach are impaired, with resulting loss of appetite. Furthermore, nicotine is habit forming and therefore makes smoking a veritable drug addiction.

Carbon monoxide binds to hemoglobin, reducing the amount of oxygen that the blood carried to the body's cells. Acrolein, ethyl alcohol, phenols, and hydrogen cyanide cause irritation of the respiratory tract, whose more serious symptoms are redness and hypertrophy of the throat and impairment of the activity of bronchial cilia. These effects in the long run can lead to chronic bronchitis.

Tar is a particularly hazardous component of cigarette smoke. It forms during the combustion of tobacco and paper in amounts that depend on the type of tobacco and paper, the smoking rate, and other factors. It has been calculated that one cigarette produces 13–35 mg of tar. In addition to various other substances, tar contains polynuclear aromatic hydrocarbons [such as benzo(*a*)pyrene, 3-methylcholanthrene, and 1,2:5,6-dibenzanthracene] which are known carcinogens.

**Table 8.2**
Death rate for smokers and nonsmokers

| Cause of death | Specific mortality rate per thousand | | |
| --- | --- | --- | --- |
| | Nonsmokers | Smokers | Pipe and cigar smokers |
| Lung cancer | 0.07 | 0.93 | 0.43 |
| Cancer of the respiratory and upper digestive tracts | 0.04 | 0.15 | 0.16 |
| Chronic bronchitis | 0.05 | 0.51 | 0.15 |
| Coronary disease without hypertension | 3.31 | 4.39 | 3.18 |

Source: P. Freour and P. Coudray, *Fumeurs en question*, Paris (1977).

Epidemiological statistics tell the story. The mortality rate, with 1 for nonsmokers, is 1.34 for smokers of up to 10 cigaretts a day, 1.7 for smokers of 10–20 cigarettes, 1.96 for smokers of 20–40 cigarettes, and 2.23 for smokers of more than 40 cigarettes a day. Table 8.2 illustrates the mortality rate of smokers and nonsmokers for various diseases.

Despite such clear evidence of the hazards of smoking, most governments have shown a regrettable insensitivity to the problem. Worse yet, instead of decreasing, the consumption of tobacco has actually increased in the last 15 years. World production of tobacco in the last decade has risen by 23%, reaching in 1975 a record amount of 5.4 million tons.

From the economic point of view, although tobacco appears to be a source of income for many countries, in reality it is a heavy burden. In Great Britain government revenues from tobacco amount to tens of millions of pounds, but the social cost of smoking (medical and pharmaceutical costs, insurance benefits, lost work days, and fire damage) has been estimated at twice that cost. In the German Federal Republic tobacco contributes 9 billion marks to government revenues, but imposes social costs of 20 billion marks.

*Alcohol*

As history tells us, the search for some chemical substance that would help man to bear his burdens, or, more optimistically, could help him to find happiness, is as old as civilization. Whether it is opium or coffee, modern psychodrugs or peyote,[c] man has always used drugs that modify behavior. Among them, alcohol has played a major role since the earliest times.

Alcohol is socially acceptable[d] and is offered as a welcoming gesture. Toasts open ceremonies, seal international agreements, and favor encounters among people. The moderate use of alcoholic beverages can be considered as a desirable social habit. Alcohol abuse, on the other hand, is a major social plague. Between these two extremes there is a whole range of intermediate stages about which judgment may be witheld but is certainly not favorable.

The alcoholic content of wine is 10–12%, that of common bear is 2–4% (as much as in wine for some beer), and that of liquor is 30–40% or more. As soon as it is ingested, alcohol starts acting with varying intensity on the mucous membranes of the mouth and the digestive tract. Once absorbed, it is metabolized to water and carbon dioxide, mainly by the liver. It takes some time for the body to eliminate the alcohol absorbed, and if the rate of absorption is greater than the rate of elimination, symptoms and inebriation may develop. Substances like aspirin, tranquilizers, and barbiturates greatly decrease the body's ability to metabolize alcohol; thus the effect of drinking under the influence of such drugs is equivalent to drinking a much larger amount of alcohol.

In large quantities alcohol can kill. Newspapers occasionally report crazy drinking bouts that end with the death of the winner. Alcohol exerts a depressant action on the nervous system. It causes sluggishness and drowsiness, impairs the ability to speak or judge correctly, and causes loss of muscular control and coordination. With the loss of self-control comes the removal of inhibitions. Since alcohol is a vasodilator, drinking may cause an acceleration of heart rate, the widening of peripheral blood vessels, and hence a feeling of warmth.

In its extreme form, excessive drinking leads to chronic alcoholism, which is a real disease and can be classed among drug addictions. The alcoholic is frequently subject to cirrhosis of the liver, gastritis, loss of appetite, and a number of psychological changes that may lead to suicide. Deprived of alcohol, the patient may be subject to violent tremors culminating in *delirium tremens*, alcoholic epilepsy, and loss of memory. Even without going to such extremes, drinking (even occasional drinking) can be quite harmful because of the synergistic effects of ethanol and because of the behavioral changes it causes, which may result in fights, falls, car accidents, and so on.

As with tobacco, what the state gets in revenue from the consumption of alcoholic beverages it must eventually repay in social costs. Hospitalization and care of alcoholics, along with all the damage caused by drinking, represent a net loss for every civilized society.

*The Military*

In an essay entitled *La vie c'est autre chose*, Gerard Bonnot writes,

Napalm is a kind of gelatin that burns, developing a heat of up to 1,000°C. But burning villages is not enough. The villagers must be properly roasted. And so the chemists set to work and soon produce a pyrogel, Napalm B, which burns at 2,000°C, sticks to the skin, and penetrates into the flesh. Then they add magnesium, thermite, which reaches 3,500°C, and some white phosphorus, which attacks the kidneys, liver, and nervous system. But men are like worms. They hide in some holes in the ground and manage to survive. And so they invent an antipersonnel bomb that spews thousands of metal barbs. But there is still a problem. Surgeons untiringly pull out barbs and heal wounds. Something new is needed. And so they develop a plastic material that is harder than steel but transparent to X rays. Now it is impossible to locate and extract the barbs.[7]

This description of some aspects of the Viet Nam War has prompted me to examine an incongruity of global proportions, an unbelievable dichotomy of the modern world that we take for granted simply because we have grown accustomed to it: Outside the official scientific community there is a world apart that spends a very large fraction of all research money (an exact estimate cannot be made), a whole industry whose enormous funds are devoted to weapons research. According to SIPRI (Stockholm International Peace Research Institute), in the 1960s $15–16.5 billion per year were spent for weapons research and development, 85% by the United States and Soviet Union, another 12–14% by France, the German Federal Republic, China, and Great Britain, and the remaining 1–3% by other nations. At present, the world spends $30 billion per year for research and development in the military sector. As SIPRI notes, the higher a nation's defense budget, the larger the fraction spent for research and development. It is also reported that in the 1950s 85% of the funds budgeted for research and development in the United States, and 77% in Great Britain, were spent by the military; the percentage decreased to 50% between 1965 and 1970. (In Italy this figure has always remained between 5 and 15%; see appendix D.)

Chemical and biochemical weapons not only exist, but sometimes they are also used, either in warfare or to quell riots. As G. Menahem writes,

Police hoses can squirt powerful jets of water as well as an icy, incapacitating fluid—a thick paste that immobilizes the rioters (and is

not totally harmless, considering that a demonstrator who had fallen
in it choked to death)—or "instant banana peel," which can turn a
crowd into a bunch of dripping sardines that the police can fish with
special nets. . . . Last, this standardization of repression is translated
into equipping the police forces of the capitalist countries with a
growing arsenal of special weapons, such as grenades that contain not
only tear gas but also CN (which irritates and causes nausea, but can
also kill if concentrated), CS (which is stronger, for it causes a burning
sensation in the throat, nose, eyes, and mouth, but was nevertheless
used against Berkeley students in 1969), and DM.[8]

These compounds were used in VietNam along with defoliants.
Agent Orange, as Nguyên Dang Tâm writes, "was employed in Viet-
Nam in amounts that no longer have anything in common with agricultural use. In the hands of American military experts, these weed
killers, these defoliants have become devastating weapons that not
only destroy vegetable resources, nutritional as well as industrial, but
jeopardize the future of the country by sterilizing its soil. . . ."[9] Not to
mention the fact, we can now add, that Agent Orange contained such
large amounts of dioxin impurities that the US Army had to destroy
its supplies in 1972 because the presence of dioxin precluded their
agricultural use. Just consider that the amounts of defoliant used in
the United States were 1.9 pounds per acre, compared with 25.6
pounds per acre in Viet Nam.

"A second example of the growing use of toxic substances in modern
weapons," adds Nguyên Dang Tâm," is CS, which has been widely
used in Viet Nam since 1964. In 1969 the US Army acquired for its
troops 2,800 tons of CS in 23 different types of munitions. . . . A special
pump was developed to blow CS powder into underground passages.
The lethal dose, 25–150 grams per minute per cubic meter, is thus
greatly exceeded. Furthermore, the victims are often weak and
undernourished—elderly people, children, and pregnant women
caught in the turmoil of a war that has been going on for more than
30 years."

Not much is known about the weapons with which modern armies
and police forces are equipped, and the little we know concerns only
half of the world. We know nothing about the other half, but we can
easily guess since the technological level attained by the Communist
bloc is comparable to that of the West in many sectors of strategic
importance. Military expenditures are also comparable, and judging
from the debates that surround disarmament talks it is evident that
neither party intends to "lag behind" the other in any field. This is
why we can speak of an incongruity of global proportions, even

though any documented evidence of chemical weapons I present concerns by necessity only those countries where information is subject to less strict controls.

Once in a while, scientific journals or newspapers talk about mysterious acronyms, botulin toxins, staphylococcus enterotoxins, and encephalitis viruses. The *New York Times* informs us that the CIA apparently spent $25 million to understand how to control the human brain and that the purpose of the study was to find ways of inducing amnesia at will by means of drugs like LSD.[10] The reporter adds that in 1975 a congressional committee established that an Army chemist, Frank Olson, had committed suicide after he had been unwittingly used as a guinea pig. We also learn that the accidental discharge of nerve gas during tests of an aerial spray tank in Utah caused the death of hundreds of sheep (other sources say more than 5,000), some of which were grazing almost 45 miles away. Most of the time all such pieces of news are sketchy and vague and raise more questions than they answer.[e]

The incongruity is evident: It truly appears that the right hand does not know what the left is doing. On the one hand, we make mighty efforts to bring to the surface the barrels of tetraethyllead aboard the *Cavtat*, a Jugoslavian ship that sank with all its cargo in as yet unexplained circumstances in the Strait of Otranto on 14 July 1974; on the other, a ship loaded with nerve gas slated for destruction is sunk, of all places, in the sea.

We do not know how many tons of toxic chemicals are stored, transported, and handled in heavily populated areas, but we impose the strictest regulations for storing oxygen or shipping packages. The whole world debates the dangers of nuclear power plants, and a few yards from our doorsteps there may be missiles armed with nuclear warheads. And we must not forget that while nuclear power plants are designed to generate electricity and built with all possible precautions, missiles are designed and built to cause the greatest possible destruction. Incidents involving nuclear weapons do happen. As stated in the 1977 SIPRI yearbook, "It is clear that accidents and incidents[f] involving nuclear weapons are frequent, occurring on a worldwide average of perhaps one every few months. There is no public record of what may be even more routine minor incidents. . . . One incontestable fact is that nuclear weapon accidents do occur, are quite frequent worldwide, and occur to probably all the different nuclear weapon systems. . . . in silos, in the air, in harbours, under the sea-surface, on land and so on" [SIPRI, *World Armaments and Disarmament*, Stockholm (1977)].

Tables included in the yearbook list 90 such accidents and incidents, one of which was the crash 10 November 1969 off the coast of
Sicily of an American plane reportedly armed with nuclear warheads
(at least in the Italian version of the story; the US government denied
it). Although the Americans and Russians are responsible for the
great majority of these episodes, the French and English have also
had their share of accidents. In addition, one cannot discount the possibility of accidents due to the unauthorized use of nuclear material.
In view of the current world situation, the sabotage or highjacking of
nuclear weapons is an eventuality that lends itself to nightmarish
speculations.

Continuing this brief survey of modern incongruities, let us consider genetic engineering. The US Senate held hearings to evaluate
the risks and benefits of this new science and Senator Edward Ken-
nedy spoke forcefully on the issue. Debates range everywhere over
the safety of a science that could be of great benefit to humanity, for
example, by developing cereal strains that could fix atmospheric nitrogen as legumes do (and therefore would not need nitrogenous fertilizers) or by producing bacterial strains that synthesize insulin and
other peptide hormones. While the official Scientific community debates the issue and research in the field is slowed down by well-
grounded safety considerations, the military in their ultrasecret
laboratories may have already used genetic engineering to develop
terrifying weapons.

To summarize: All our efforts to prevent accidents, all our controls,
precautions, and toxicological tests concern one side of human activity. There is another side to which all this does not apply, a side that
we occasionally glimpse when we read in the newspapers that
*thousands* of sheep were killed by nerve gas or that *thousands of tons* of
lethal weapons are going to be destroyed.

In other words, it is a sure thing that every scientific study is being
exploited for its *destructive* applications. Furthermore, in general the
most important scientific advances are possible only because research
has already been done for military purposes. When it comes to applying these scientific advances to the quality of life, *then* we start
questioning and debating. The good is only a possibility; evil is a certainty. Anyone who wishes to give a meaningful content to the expression "new patterns of development" would do well to keep this in
mind.

On the whole, then, it is a most incongruous situation. From habit,
or conformity, or misinformation, we accept very serious risks, some-

times in exchange for insignificant benefits. We accept smoking, alcohol, and poisoning from exhaust fumes. For centuries we have accepted an unrestrained military development in spite of the fact that we have all suffered its consequences. Given this absurd background, the disenchanted observer can only remark that in their crusades against products that are essentially harmless in an absolute sense (let alone relative) our institutions behave like fire fighters racing through a burning town to snuff out a match.

# 9

## International Efforts to Curb Risks

At first international bodies sought to remedy only the problems posed by chemical technology that had aroused public opinion. But for some years now experts at OECD, EEC, and WHO have been at work on directives using a scale of priorities that endeavors not only to specify the most serious problems but to suggest corrective measures that are technically feasible, bearing in mind the need to harmonize the legislations of the various countries and to prevent inconsistencies in international trade. In addition, they are taking into account the cost of remedial measures, and efforts are made to avoid imposing such strict limitations that they would require expenditures disproportionate to the goal; limited funds make it desirable not to waste money on futile enterprises. Also, implicitly, or explicitly—as in the case of the US government's Toxic Substances Control Act, or TOSCA for short—the utopian concept of "zero risk" has been replaced by that of "reasonable risk." Efforts are also made at the legislative level to consider risks along with benefits. Last, careful consideration is given to the risk of *not acting*, which means that the risk of an action is weighed against the risk of no action (using an insecticide or tolerating malaria, for example).

### The Concept of Risk

It has been argued[1] that society and the individual have accepted the fact that risk is a normal aspect of our life. There are risks that are taken voluntarily for all sorts of reasons, from profit to sport, in return for an expected benefit. Such risks are taken the more willingly and frequently, the higher "the individual's perception of his own ability to control the risk-creating situation." But there is a marked change in attitude "when the individual no longer believes he can control his risk exposure." This occurs when the risk, is involuntary, that is, when it is no longer the individual but a societal group that controls the risks-benefits balance. Many of the modern technical systems create such involuntary risk exposures—for example, transportation systems, public utilities, and food supply systems. "Under these circumstances, the individual exposed to an involuntary risk is fearful of the consequences, makes risk aversion his goal, and therefore demands a level of such involuntary risk exposure as much as one thousand times less than would be acceptable on a voluntary basis."

This argument is fully supported by, on the one hand, the attitude of the individual toward such voluntary risks as smoking, alcohol, and drug abuse and, on the other hand, by the reaction of public opinion to environmental hazards that gives marginal problems a disproportionate amount of attention.

To strike the right balance between technological progress and environmental problems, decisions must be based on two kinds of evaluation. One is the assessment of the risk associated with a particular technological option: This is difficult because of the number and diversity of the chemical, biological, and environmental agents that may be hazardous to man and because of the complexity of both the agents' modes of action and the organisms that may be affected. The other is the evaluation of the level of risk that may be considered acceptable: This involves a value judgment and is therefore uncertain in an absolute sense. For example, when we say that a certain substance can cause cancer in one person in a million, the point of view is quite different, *a posteriori,* depending on whether we happen to be that person or one of the other 999,999. Furthermore, it is not always clear who should decide which risks are acceptable for whom, in what terms, and why. It is from this question that arise some of the cardinal issues of our time, such as the role of the elites (including the technological elite), the view of justice, and the problem of the fair distribution of risks, costs, and benefits.

Assessing the level of risk associated with a certain agent, whether chemical or physical, is a very difficult enterprise. Risk assessment is relatively straightforward when specific damage can be positively traced to a given substance (as in the case of thalidomide, for example), but it becomes extremely difficult when one is dealing with the potential, long-term detrimental effects of exposure to a product of recognized usefulness (an insecticide, for example). As A. Plant remarks, "Our toxicological testing too often raises more questions than it answers. . . . I suspect that we often operate on an overkill procedure and ban materials that really present no danger in their normal use."[2]

When evidence emerges of the potential dangers of a particular agent we are confronted with a difficult decision. One approach is to ban the product immediately, regardless of its benefits, the availability of substitutes, or the socioeconomic consequences. The opposite approach is to defer action until the level of risk can be assessed with ad hoc documentation. In this case one runs the risk of intervening too late and making the damage worse. The regulatory agencies, and hence the public, are continually torn between these two extremes.

Very few people are aware of the thousands of man-hours, the number of test animals, the amount of equipment and money that are required for toxicological testing of just one substance, and very few people realize that even the most painstaking research may not yield definite results. To give an example, "Although $2 billion and 30 years have been spent studying the biological effects of ionizing radiation, acceptable exposure levels are still debated."[3]

The effects of chemical compounds on humans are just as difficult to determine. Because of the lack of accurate information, once a potentially harmful agent has been recognized there is a tendency to demand its immediate ban and to reduce the risk to zero before its biological effects can be clearly ascertained; this frequently entails prohibitive costs without commensurate benefits. Furthermore, the consequences that the ban of a chemical may have on the environment, public health, or the economy are not always given proper consideration. Representative of this point of view is Borlaug's bitter polemic against opponents of agrochemical products (see chapter 2).

*The Role of the OECD*

The legislative actions taken within the industrialized world have been conceived and coordinated by the OECD. At first, the Environmental Committee of the OECD concerned itself with the problems related to air and water pollution through two ad hoc subcommittees. Subsequently (1971), a third subcommittee was created for the purpose of controlling the accidental diffusion of chemicals in the environment. Although a later addition, this subcommittee deals with problems that are certainly as relevant as those debated by the other two. This may be taken as evidence of a fact I have often remarked upon, namely, the tardiness of international action.

In the beginning, this subcommittee followed the general trend by concerning itself with mercury, cadmium, and polychlorinated biphenyls. Later on, however, it attacked the problem from a more general point of view and formulated the criteria that form the basis of current regulations. Implicit in such criteria, and in the laws inspired by them, is the concept of the *provisional* and often *questionable* nature of the criteria themselves, particularly with regard to the definition of harmfulness.

A number of nations have enacted legislation to control the production, marketing, diffusion, and disposal of chemical products. The common object of all such laws is to entrust government agencies with the evaluation of the potential effects of chemicals and the power to act before damage is done. With regard to their formulation, these laws can be divided into three groups. A first type of legislation is inspired by the general principles of Anglo-Saxon law. Producers and distributors of new chemical compounds are not required to send the government any premarket notification, but take full responsibility for evaluating such effects as toxicity, persistence, accumulation, and so forth. Norwegian and British laws are based on these criteria. A second type is the Canadian law, which is based on more fiscally-minded criteria. As a rule, it does not require premarket notification from producers or distributors. At the same time, it gives the government the right to ask for whatever information is needed to prevent the marketing of toxic substances. Notification is mandatory, however, when a product's annual sales are expected to be over 500 kilograms (1,100 pounds). On the basis of information provided by the manufacturers or gathered in-house, the government can then impose regulations, controls, and penalties. The third type of law, adopted by Switzerland, Japan, and the United States, requires manufacturers or distributors to submit a premarket notification supplemented by information concerning the chemical and physical properties of the new product and its biological effects. On their part, the regulatory agencies are required to examine the documentation and decide whether the product should be marketed and under what conditions.

Clearly, the Swiss, Japanese and US laws entail very active government intervention in the licensing of new products. In Japan, for example, once notification has been received, a product must be subjected to official testing to assess its persistence, accumulation, and toxicity to man. After sufficient data have been gathered, the chemical is classified in one of two categories: If in any way hazardous, it is classified as "specified substance" and can be used only in conditions determined by a special decree; if safe, it is allowed to be marketed freely.

The Swiss law has chosen $LD_{50}$ (average lethal dose) as the toxicity index of a substance. Anyone who wishes to manufacture, import, sell, or dump a particular chemical must notify the authorities; depending on its $LD_{50}$, the chemical is classified in one of five categories, each corresponding to precise regulations. Government agencies take part in the evaluation and classification of all chemicals and see to it that regulations are enforced.

By far the greatest impact on chemical production and international trade, at least in a relative sense, is expected to come from the Toxic Substances Control Act (TOSCA) that became effective in the United States on 1 January 1977. According to the new law, anyone who wants to manufacture or import into the United States a new product or apply an old product to a new use must notify the EPA 90 days in advance. The agency has three months to make a judgment. After the term expires, if no objections have been raised by the EPA the product can be manufactured and/or marketed. The distinction between new and old products is based on an inventory of all chemicals known as of 31 December 1976. This list, which includes more than 31,000 chemicals, has been compiled with the collaboration of manufacturers, distributors, academics, and so on. After receiving premarket notification and related documentation, the EPA can request an adjournment, in whose support it must specify matters on which it requires elucidation or additional testing. Depending on the conclusion it reaches, the EPA can prohibit some uses, set labeling regulations, and establish emergency measures.

Beside controlling new products, TOSCA requires that proper attention be paid to chemicals that have been freely marketed in the past and in some cases have achieved widespread diffusion. These chemicals are ordered on a scale of priorities that is based on annual production levels, toxicological tests, environmental persistence, and uses. Every year, starting in 1977, 50 chemicals are thoroughly scrutinized. Heading the list are chemicals such as benzene, asbestos, arsenic-based compounds, alkyl oxides, and vinyl chloride. The fact that their health hazards are being considered only now after decades and even centuries of use, is worth some thought.

## The Role of the EEC

Legislative action by the Community dates back to 1967. The directive of 27 June 1967, which was enacted in all member countries in 1972, concerned hazardous chemicals and set labeling and packaging regulations. It soon became evident that the directive was inadequate to cope with the serious damage caused to man and his environment by the growing diffusion of unsafe substances. A number of modifications was therefore made to the initial directive, as a result of which PCBs were banned, the use of vinyl chloride as an aerosol propellant was prohibited, and restrictions were imposed on other products.

The sixth amendment to the 1967 directive is based on the criteria that the hazardous properties of a given substance must be known *prior* to its introduction into the market and that the substance's commercial development, both qualitative and quantitative, must be monitored so that its impact on man and the environment may be assessed.[4] These are the main points of the new amendment: The 1967 directives defined eight categories of hazardous substances—explosive, combustible, easily flammable, toxic, harmful, corrosive, and irritating; the new amendment adds two more categories—substances hazardous to the environment and carcinogens. In addition, a group of '"highly toxic substances" is identified within the category of toxic substances. Premarket notification is made mandatory for all new chemical products. Notification must be supplemented by back-up documentation containing data obtained according to specified objective parameters. The member countries are expected to designate the agency responsible for implementation of the directive and to set up a mechanism whereby any regulation imposed on a particular chemical in one country can be implemented in all countries. The documentation must contain sufficient data to identify the substance and determine its degree of purity, as well as information concerning methods of detection, uses, expected production or import levels, handling and shipping precautions, and first-aid measures. Parameters are specified for chemical-physical data, acute toxicity data for different routes of administration, mutagenic and subacute toxicity tests, ichthyotoxicological and ecotoxicological data, as well as information regarding persistence and methods of destroying the substance both in bulk and when dispersed. Additional parameters concern the mutagenic, teratogenic, carcinogenic, and bioaccumulative properties of the compound.

*Consequences of the International Actions*

The laws that have been enacted in the last few years to control water and air pollution and the production and spread of chemicals are going to impose a heavy burden on both government and industry. The government is faced with the problem of acquiring extensive equipment and the skilled personnel needed to evaluate applications for industrial waste disposal, regulate the marketing of new products, and carry out the necessary controls. It is one thing to set acceptable mercury levels in industrial effluents at 0.001 ppm (parts per million) and another to be able to enforce the regulations. It is one thing to request detailed documentation from the makers of a new product and another to have the technical personnel that will read and evaluate each application. We are dealing with voluminous documentation that requires expertise in toxicology, industrial chemistry, and ecotoxicology and that tends to pile up in government offices.

Reports by the Swiss OECD representatives on the implementation in their country of the legislation regarding toxic-substances control clearly show that estimates of the number of technical people required for granting licenses and exercising controls were woefully unrealistic. Additional technicians have had to be hired, and naturally costs have increased as well. The EPA estimates that the implementation of TOSCA will require a department of more than 1,000 people.

Industry will also be facing very large expenditures. Carrying out a complete project of data gathering and evaluation for a new product in accordance with the new regulations requires expert personnel, laboratories, and money. The technical staff must be sufficiently skilled to appreciate all the biological, chemical, and physical data on which the evaluation of a product is based. This requires knowledge of toxicology, pharmacology, physiology, biochemistry, pathology, organic chemistry, and more.

The necessary laboratory equipment is complex and costly. Toxicological testing requires extensive facilities for animal tests, including equipment for environmental conditioning, strain selection, data processing, and so on. It is estimated that 50 professionals, 50 technical assistants, and 5,500 cubic yards of space are needed for a complete toxicological laboratory. An OECD study estimates at $25,000 (1971 value) the cost of testing a biologically inactive chemical (that is, a chemical that is not a drug, biocide, etc.), not including the cost of carcinogenicity tests. The cost of testing a biologically active chemical is estimated at $600,000–700,000, plus $20,000–30,000 for carcinogenicity tests.

There is no general agreement on these figures, however. At the 1975–1976 hearings that preceded the approval of TOSCA, Dow Chemical experts testified that the cost of testing a product would amount to $500,000. At present, only multinational pharmaceutical firms, and a few national companies, can afford such costly research. As for nonpharmaceutical firms, there are only three in the United States that can afford it. Smaller firms are making plans or joining together to acquire the needed facilities. The annual cost of implementing TOSCA to the whole US chemical industry will be $80 million according to the EPA and $2 billion according to Dow Chemical.

The new legislation will force the chemical industry to expand its facilities and face increased costs. It will also bring about considerable changes in its research and marketing policies. One worry voiced by US chemical manufacturers concerns the question of confidentiality. To comply with TOSCA, they will have to supply the EPA with data that are normally considered trade secrets, such as product specifications, production estimates, intended uses, potential biological effects, and persistence. "The best way to make your proprietary information public is to send it to Washington," comments an anonymous industrialist interviewed by *Chemical and Engineering News*.[5] Once such information is made public it becomes available as well to foreign competitors, who might use it to develop the product and capture the market.

Another cause for concern is the duration of patents. Evaluation by the EPA of the information contained in premarket notifications might take years. This means a delay in marketing a new product from the patenting date and, given the limited validity of a patent, a reduction in the period of its commercial exploitation. Many experts expect a decline in the research and development of new products similar to that observed in the pharmaceutical sector (see chapter 3). There is a distinct possibility that after developing a new product, and spending large sums of money for the required tests, the manufacturer will not be granted a license. As a result, as Union Carbide's Browning remarks, "A lot of people are going to make the decision along the lines of process improvement to sell the same product." Efforts will be directed at finding new uses and new markets for old products and at improving production systems.

One factor that is gaining in importance is the purity of a product. In the case of trichlorophenol-based herbicides, for example, the fact that they normally contain dioxin impurities has effectively determined their disappearance from many markets. It is expected that many sectors of industrial chemistry will reflect this new outlook and that quality, much more so than in the past, will be the determinant factor from the commercial standpoint.

Small- and medium-size firms, at which the level of innovation is often relatively high, will experience the most difficulties. Besides creating problems insofar as confidentiality and licenses are concerned, the new legislation will force such firms to acquire expensive facilities for biological testing that they can ill afford.

In the view of many observers, however, all the difficulties caused by the need to upgrade testing facilities are more than compensated by the benefits expected from the new safety regulations. As J. R. Leach, head of the safety management program at the National Institute of Health in 1976, remarks, "No one will deny that safety reduces pain and suffering and, therefore, has great humanitarian value. But let's not kid ourselves. Safety also saves employers large amounts of money and makes damn good business sense."[6]

If we set aside value judgments and concentrate instead on facts and figures, we find ample evidence to support Leach's statement in many reports by US health agencies. If it is true, as generally accepted, that 60–90% of cancer cases are due to environmental factors, then cancer can be prevented to a considerable extent by limiting the spread of carcinogens in the environment. According to the 1975 US report on environmental quality, in 1974 cancer killed 358,400 people in the United States. Approximately 1 million people are undergoing treatment; 900,000 new cases of cancer are diagnosed every year, and while 300,000 of them involve the skin and generally do not pose a grave threat, the others, as everyone knows, are very serious. The economic and social impact of cancer is enormous. In the United States alone almost $2 billion are spent every year on just the hospitalization of patients. Additional expenses, drugs, and dotor fees raise this figure to tens of billions. One should also add about 1.8 million work-years lost to the national economy because of the unemployment or underemployment of cancer victims. All this demonstrates, without emotional appeals, that the cost of prevention is substantially lower than the cost of a cure.

Analogous considerations are advanced in the field of occupational diseases and accidents—a field that in part overlaps the preceding one. According to the National Safety Council, the total US cost of on-the-job disabling injuries in 1974 was $13.6 billion. Since 2.3 million disabling injuries were reported, the cost of the average injury can be calculated at $5,900. Financial responsibility for the health problems caused by exposure to chemicals is a very heavy burden on the chemical industry. According to reports compiled by American industrial associations, a growing number of very large claims are been filed by victims, with the courts often awarding huge judgments. As a result, insurance coverage for this type of risk is becoming prohibitive. When policies come up for renewal, premiums frequently go up by a factor of 30 or 50, and many companies refuse to sign the new policies.

This insurance problem is becoming even greater than that of malpractice. To give an idea of the figures involved, in 1975 premiums paid by doctors to insure themselves against malpractice suits amounted to $750 million; by a very conservative estimate, insurance premiums for coverage of on-the-job exposure to chemicals raise this figure to $16 billion for the same year. With every lost suit, the economic burden has become heavier; the average award rose from $11,644 in 1965 to $79,940 in 1973. (And the legislative criterion for attributing responsibility has also changed. Whereas once it was the seller who was prosecuted for the damages caused by the product, now it is the manufacturer who is called to answer.)

# 10

## Concluding Remarks

"Attainment of the middle-class style of life," writes Nathan Keyfitz, "is what constitutes development in countries as widely separated geographically and ideologically as Brazil and the U.S.S.R. In the process peasants gain education, move to cities and adopt urban occupations and urban patterns of expenditure. . . . These changes can be visualized in terms of a definable line, comparable to the poverty line officially drawn in the U.S., across which people aspire to move."[1,a] According to Keyfitz, the problem is not so much to know how many people live on earth, but rather how much they consume by the middle-class standards of living, the point being that while world population is growing, the standard of living of the so-called middle class is also rising on an absolute scale: "Production of most things consumed by the world's people has been increasing at a higher rate than the 1.9 percent per year of population. During the period from 1960 to 1963 meat output increased at 2.8 percent a year, newsprint at 3 7 percent, motor vehicles at 6.8 percent, and energy consumption at 4.9 percent."

This is the general trend in the world today. In this context it may be interesting to report what occurred at the Conference on the Human Environment held in Stockholm in June 1972. The efforts of the industrial nations at limiting pollution were interpreted by the representatives of many developing countries as an attempt to curb the development of the Third World. To quote Mrs. Gandhi:

Aren't really poverty and need the greatest pollutants? . . . How can we talk to people who live in villages and slums about keeping oceans, rivers, and air clean when their own lives are polluted from the start? The environment cannot be improved in conditions of poverty. Nor can poverty be eradicated without the use of science and technology. . . . [With regard to overpopulation] we can see that, as far as pollution and waste of resources are concerned, a population increase of one in the affluent countries, at that standard of living, is equivalent to many Asians, Africans or Latin Americans at their current level. . . . It would be a great irony if the fight against pollution should turn into another affair, with a few companies, enterprises, or nations profiting at the expense of many."

On his part the Cuban delegate Pelegrin Torres stated, "If a farmer from the northeast of Brazil could choose between his average life span of 30 years without pollution and a life of 70 years in countries choked by smog, he would certainly choose the latter."

To deal with the current crisis, the West speaks of nations that must "act as locomotives," advocates "stimulating the economy," and looks for alternative energy sources. The fact is that our system views with dismay a growth rate of "only" 3% and rejoices instead when the GNP of a nation returns to a "healthy" 5–7%. As reported by *L'observateur de l'OCDE,* this is what the governments of the member countries declared in June 1977: "The cooperation for economic development must consent to the realization of a twofold objective, which consists in increasing revenues and meeting man's essential needs in all the developing countries."[2] The article goes on to note that productive investment is the cardinal stone of this program and technology the engine that allows the increase in productivity to accelerate. It also remarks with satisfaction that "in particular, we now have at our disposal a high-yield agricultural technolgy—in the guise of improved practices, new cereal strains and fertilizers—that favors small family enterprises. . . ." In other words, the system has an inherent rigidity that, judging from what is happening, seems to favor neither a voluntary return to a lower standard of living nor a drastic change of attitude on the part of the average man.

Exhortations to save energy go largely unheeded. A survey conducted in the United States will suffice to prove the point. In a dramatic message President Carter asked the nation to reduce consumption voluntarily (that is, without recourse to price increases, controls and penalties) by setting thermostats at 65°F (18°C) during the day and at 55°F (13°C) at night. A subsequent telephone survey revealed that most Americans kept their daytime temperature at 66°F (19°C) and nightime temperature at 64°F (18°C). Inspection showed actual temperatures of 70°F (21°C) and 69°F (20.5°C), respectively. In effect, any measure that threatens even minimally our standard of living provokes immediate protests, be it bicycling on Sundays instead of driving, observing speed limits, or closing butcher's shops two days a week.

There is a great deal of talk about "new patterns of development," but nobody has been able to define, let alone discover, them.[b] We often hear voices advocating a return to a patriarchal type of agriculture as the best possible system from the ecological point of view, but these messages go largely unheeded, too, even in periods of crisis. Perhaps this is because the people who preach this approach are in general firmly ensconced in the urban middle class and quite unwilling to do it themselves. It is the farmers who are asked to devote their lives to a type of agriculture that requires back-breaking labor and offers in return a meager and uncertain income.

There is something schizophrenic about the attitude of modern man. What people say is entirely different from what they do. By and large, everyone agrees on the sacrifices that *others* ought to make. It is said that man is the only individual-oriented animal among species-oriented animals; except for brief periods when dramatic events engender great human solidarity, the individualistic mentality always prevails over the good of the species. The individual normally behaves toward society like the man who attributes urban traffic jams to the selfishness and lack of discipline of the citizens (or rather, of his fellow citizens), but is always ready to double park "a moment" to go buy a pack of cigarettes.[c]

If man behaved as he is supposed to by accepted codes of ethics, many problems would not be so serious and could even be solved. Hunger in the Third World could be alleviated if the surplus of pears, apples, and tomatoes were shipped there instead of being destroyed by bulldozers. Instead, for a whole slate of reasons that generally have to do with a conflict between the interest of the community and the interests of the individual or a group, goods are destroyed in one place while a few hours' flying time away people starve.[d] Although this is generally thought to be a sad exception to a hypothetical code of altruistic behavior, it is in reality the rule. As a rule, we make tanks, go to the moon, and buy fur coats rather than share our wealth with the poor. It is a fact, and we had better recognize it.[e] Thus, when we write economic theories or draw up plans for the immediate future we had better program into the computer man as he really is.

Judging whether there is a concrete possibility that we shall be able to formulate a wise policy that will be followed with conscious determination by most of mankind is something that depends on the knowledge, perception of reality, clarity of vision, and optimism or pessimism of each of us.[f] In any event, it seems fair to predict that it will be a slow process and that its success will depend not only on technical solutions but above all on the development of moral education and civic responsibility, both of which are virtues that must be patiently cultivated in every member of the human community.

In the meantime the world must go on. In order to go on, and to avoid economic crises more serious than the current one, our productive systems will remain the same in their general lines. They may be improved—and we hope they will be—so as to take into account our environmental problems, but it is clear that in the next decade agriculture will follow modern industrial patterns, metals will continue to be used, plastics will be the basis of new materials, and people will eat preserved, precooked, or industrially processed foods. It is clear, in other words, that we shall need fertilizers, energy, ethylene, and pesticides; that we shall use oral contraceptives for some time to come; that we shall go to hairdressers and use more cosmetics than ever—in brief, we shall continue to use up essentials for purposes that may or may not be essential. Furthermore, we shall continue to live much as before; do not expect a spontaneous return to the patriarchal type of agriculture that people have in fact spontaneously abandoned because of its hard work and uncertain returns or from a wish to live in the city.

What is the meaning of all this? It means that our current productive systems, good or bad, will remain for a while longer. And that we shall have to generate enough energy to produce the food and materials and to power the transportation systems required by these systems. More important, it means that there will continue to be a close correlation between production and employment, that the GNP, along with the balance of payments (which, for many countries, depends on arms sales), will continue to be a leading economic indicator, and that industrial production will continue to require raw materials, a labor force, and technology. Unless, of course, a superior being is able to come up with a new pattern of development and to impose it on the whole world democratically, quickly and successfully—unless a miracle happens, in other words. But those who do not believe in miracles are left with a hard task.

## Public Opinion

In all democratically governed countries individual participation in the decision making process has greatly increased in the last 30 years. It occurs in two ways—by the election of representatives and by the expression of opinions through the press, demonstrations, referendums, and strikes. Public opinion has a peculiar influence on the issues we have discussed since choices in this area are no longer a matter of majority vote but depend, at least in practical terms, on the *veto* power of a minority. It is not surprising that afterthoughts, debates, and revisions should mark every stage of the development of such public works as thermoelectric or nuclear plants or the licensing of products for mass consumption. But beyond the specific disagreements involved in these matters is the conflict between community interests and special interests. Plans for the development of nuclear power are approved by the great majority of our representatives, but when it comes time to laying the first stone of the first plant, they are blocked by the veto power of the local opposition. Rightly or wrongly, depending on the circumstances (and it is often hard to judge whether it is right or wrong), a government's or a majority's "power to act" must contend with the "power of not to act," a form of negative power that is assuming a leading role in our democratic societies and can influence for better or worse the economic and social life of a nation.

History continually reminds us that there are ways of undemocratically imposing on the masses life styles that disregard the "development of moral education and civic responsibility." The tragic story of the Kulaks in the early days of Soviet industrialization is a case in point. As for the influence of Russian public opinion on important technological developments, it is instructive to read what *Chemical and Engineering News* for 6 November 1978 reports about Soviet nuclear reactors: "The Russians are noncommittal about where they store their wastes, where they do their enrichment. . . . As elsewhere, spent fuel is stored in water near the reactors until shipment. Where it is shipped, nobody claims to know, saying only that it is sent away by rail. . . . The Soviet attitude toward safety is not as stringent as that in the West."

Let us entertain only one future scenario, namely, things will be done the democratic way and with respect for the rights and freedom of man. This calls to mind what biologist E. Mayr said during a conversation with Dobzhansky and Montalenti at the International Colloquium on the Origin of Man held at the Accademia dei Lincei (Rome, 28–30 October 1971) and reported in the academy's minutes of 1973:

The difficulty with modern man is that the freedoms which we cherish the most are freedoms that came in the period of Enlightenment," when our modern human society rebelled against feudalism, rebelled against special privileges, rebelled against all sorts of establishments, and replaced them by the freedoms of *égalité, fraternité,* and *liberté*. Now, 200 years later, the question is whether we, as a social animal, have not only the right, but—and I raise this merely as a challenging question—whether we do not also have the obligation to challenge some of these freedoms. For instance, in the United States now, there are very many people who say that the freedom to throw as many children into this world as one wants and inflict them on the rest of society is no longer a legitimate freedom. . . . Many other freedoms are now being questioned, such as the movement of people. Perhaps people should not have the freedom to go anywhere they please, if they affect society at another place adversely. . . . Now, the whole modern technology, the affluence, the population explosion, the new density of populations raise problems that did not exist in 1786 or 1789. And it is perhaps necessary that we look at all value systems very carefully in the light of the changed conditions. We don't have to become Fascists; we don't have to become totalitarians; we don't have to become Marxists; but I think, as thinking people, as people who have a feeling of responsibility, we must realize that our generation determines a great deal about the future of mankind. I think, in this role we have a deep obligation to look very hard at the accepted values that are more than 200 years old and examine them in order to find out whether they are still suitable for the year 1971.

The issue raised by Mayr is more than pertinent to the theme of this book.

## The Need for Prevention

All the efforts of medical science are aimed at alleviating man's suffering and delaying his death. To a certain extent, suffering must be ascribed to fate. But a great deal of it, as we have seen, is due to man's own behavior: to his environment; to the food he eats and the drugs he takes; to smoking, drinking, and narcotics. This part of it, at least in theory, could be prevented through individual and community action. "Most individuals," writes J. H. Knowles, in 1977 president of the Rockefeller Foundation, "do not worry about their health until they lose it. Uncertain attempts at healthy living may be thwarted by the temptations of a culture whose economy depends on high production and high consumption. Asceticism is reserved for hairshirted clerics and constipated cranks, and every time one of them dies at the age of 50, the hedonist smiles, inhales deeply, and takes another drink."[3] At the individual level, prevention of disease means forsaking bad habits; at the community level, it means making efforts to reduce the health hazards in our environment.

In the United States there is a current of opinion strongly in favor of preventive care as a remedy for the heavy economic burden of national health care. As Knowles remarks, "The cost of private excess is now a national, not an individual, responsibility. This is justified as individual freedom—but one man's freedom in health is another man's shackle in taxes and insurance premiums." As a contributor to *Science* concisely puts it, "Uncle Sam wants you to be healthy and inexpensive."[4] Similar considerations (see chapter 2) apply to environmental pollution and exposure to toxic substances. We have two means of coming close to a solution of these problems—information and research.

## *The Importance of Information*

Information is essential to democracy, particularly when the bond of trust between the governed and the governing bodies is marred by the suspicion—sometimes justified—that choices officially ascribed to technical reasons may in fact be political. Technologically advanced nations are confronted with very important decisions with regard to energy, chemical production, and nutrition. It is essential that both the governed and the governing bodies should know very well what they are talking about. (In the first months of 1974, right after the beginning of the energy crisis, it became apparent that a great many people in Italy, including some in positions of great responsibility, believed that oil imports were needed solely to fuel motor vehicles.) Choices have to be made, and we should all understand both the risks and benefits expected by choosing action as well as the consequences of choosing inaction.

Citizens need information to understand common products: that plastics come from petrochemistry; that processed food does not go bad because special additives prevent spoilage of fatty substances; that electric energy comes from the most part from the combustion of oil; that agricultural production requires a substantial expenditure of energy. They need information to distinguish between consumption and consumerism, between the soap for personal hygiene and the mineral water in expensive spray cans that many ladies of our bourgeoisie, dazzled by publicity wizards, spray on their faces. They need information to evaluate our options for economic development, and first of all our choice of an energy policy. (We know a great deal about nuclear power, but not nearly enough about the drawbacks of the other energy sources: "Over the past 20 or 30 years the federal government has invested well over \$1 billion attempting to measure the public health costs associated with nuclear power, and until recently almost nothing was done to measure similar hazards of fossil fuel power—in retrospect, a scandalous omission."[5])

It is important to know what contribution solar or hydroelectric power can make to our energy needs in the immediate future, how many dams collapse every year, and the human cost of extracting and burning fossil fuels. It is important not to overestimate some potential energy sources or underestimate others. Above all, when choices have to be made between different alternatives, we must fully understand their respective risks and benefits; only in this manner can we avoid the mistake of suspecting new technologies while at the same time accepting old technologies that are deemed safe only because familiar.

Information can help to prevent jarring inconsistencies, for example, that cigarettes are sold to youths 13 years old while diabetics are deprived of saccharin and that people ignore the proliferation of nuclear weapons while stopping the construction of nuclear power plants. Information is needed now, and will be needed even more in the future, to save people from falling into advertising traps and the marketing strategies exploiting the considerable margins of error in the operational definitions of toxicity, unsafety, and synergism. Information can prevent a great fuss being made over trivia, while momentous problems are ignored. And information is the necessary basis on which to train personnel in sufficient numbers and with sufficient skill to deal with such difficult problems. The present lack of such trained personnel is a serious problem.

## The Importance of Research

Research offers the hope of remedies for the damage that has been done and new technologies with fewer drawbacks than the old. Unfortunately, the almost unlimited supply of cheap oil discouraged research into renewable sources of energy and organic raw materials. Wood—a material whose supply is enormous and whose continual renewal can be ensured with proper management of the resources—is only one example of the alternatives that have been overlooked. Now that the crisis is here, we scramble around for solutions hurriedly and haphazardly, by reconsidering all possible alternatives at the same time. It may be obvious in retrospect, but no less true, that had we spent a little less money for petroleum research and a little more for research on energy alternatives from 1946 to 1974, today our options would be much clearer.

Research on quick and inexpensive ways of testing chemicals for long-term toxicity and carcinogenicity has also been neglected. At present such tests are generally performed on animals. They are time consuming and costly and require great care to guarantee absolute constancy of the parameters (such as the diet fed to control groups and test animals) and to reveal the smallest evidence of cause-and-effect relation. Also, any more extensive use of animals for toxicological testing would sooner or later face opposition from the various humane societies.

Testing thousands of chemicals for long-term effects by conventional methods is definitely out of the question. However, faster and cheaper tests have been under study for some time. Performed on microorganisms and animal cells *in vitro*, these tests, the best known of which is the Ames test, give useful indications about the mutagenic[g] or toxic properties of chemical substances. Their time scales and costs are considerably smaller than for animal tests.

Research in this field is very promising and may eventually give us one solution to our problems. Even more important is the fact that more and more researchers are aware of the need to expand their point of view to take into account the global picture, instead of concentrating on short-term, partial answers. A case in point is the vigorous research that is being carried out on insect hormones in the hope of finding selective insectides nontoxic to higher organisms. Other examples are research in solar energy, the photovoltaic effect, biocompatible materials, and slow-release fertilizers, as well as the growing use in industrial chemistry of enzyme systems as substitutes for chemical reagents and solvents. Last, genetic engineering may make great advances in the field of biological nitrogen fixation, thus completely solving the problem of fertilizers.

Clearly, the world is at a turning point. On the one hand, we have a constantly growing population and, in particular, a growing middle class that takes a lion's share of our resources. On the other, we have a shortage of resources and dangers inherent in our production systems. We must find a way to reconcile expectations and production, costs and revenues, society and the individual, risks and benefits. It will not be an easy task; we should confront it realistically. It is better, perhaps, not to indulge in facile optimism and put too much reliance on human wisdom or the ability of national and international organizations to solve our problems.[h] The route we have embarked upon appears to be a one-way street: with the possible exception of communist China (whose economic policies suggest *rapprochement* with Western ideas), the word "progress" has the same meaning everywhere.

We may draw these conclusions from all that has been said: For at least the foreseeable future the world will be traveling the same no-return route; and chemistry's role will remain fundamental. But chemistry is not devoid of risks. In my opinion, man must understand and accept that there are no benefits without risks. I recognize that this kind of statement is extremely dangerous, in that it may be used to justify ill-conceived actions. I should emphasize, therefore, so that this statement may be meaningful from the practical as well as the ethical point of view, that every effort must be made at every step to maximize the benefits and minimize the risks. In practice, this is a difficult goal to achieve in view of the uncertainties and ambiguities that surround all the issues I have discussed. Nothing is obvious, straightforward, or clear-cut for the decisionmaker. If we then add individual selfishness and special interest groups that confuse the issues and use events or news for their own purposes, the task becomes next to impossible. The only instruments at our disposal for reaching agreement and success are research, information, schools and education.

This is the greatest challenge of our time and the ground whereon progress will fight out its battles for survival. The outcome is not easy to predict. For people who are engaged in science and technology the current situation poses a serious warning—the same warning that E. Schrödinger gave scientists several decades ago in *Science and Humanism* perhaps sensing what the future would bring: "Never lose sight of the role of your particular discipline in the great course of the tragicomedy of human life; keep in touch with life, not so much with practical life, but rather with life's fundamental ideals; and let life keep in touch with you. If you do not, in the long run no matter what you are said to have done, your work will have been in vain." The clamor of war and then the striving for prosperity have deadened the sound of his voice. The time has come for all who do science to hear it again.

# Appendix A    Food Additives

Food additives fall into the following principal categories:

*Antimicrobial and Antioxidant Agents*

*Acetic acid (E 260)*    Recommended for making dough and leavened pastries up to 4%; also used in combination with sodium acetate; naturally found in wine; no known drawbacks

*Sodium ascorbate*    Sodium salt of ascorbic acid (vitamic C); used as an antioxidant in food processing and in the preparation of frozen fruit and vegetables; no toxic effects; recommended dose 2.5–7.5 mg/kg; at 5 mg/kg has diuretic effects

*Ascorbic acid (vitamin C)*    Antioxidant used in grain and oil emulsions; inhibits browning of fruit and potatoes

*Ascorbyl palmitate*    Antioxidant used in oils (except olive oil) and fats; delays growth in rats when fed as 5% of the diet, but protracted feedings of diets containing 1–5% of ascorbyl palmitate do not have any noticeable effect

*Benzoic acid (E 210)*    Antimicrobial agent particularly effective against molds; very low toxicity

*Sodium benzoate (E 211)*    Has low toxicity when used as an antiseptic

*Methyl-, ethyl-, propylparaben*    Antimicrobial and antiseptic preservatives; recommended dose up to 2 mg/kg

*Butylated hydroxytoluene (BHT)*    Antioxidant used in fats and oils; has some drawbacks at levels higher than 0.5%

*Butylated hydroxyanisole (BHA)*    Antioxidant used in fats and oils; some reservations have been expressed about its safety

*Boric acid*    Antimicrobial and antiseptic preservative for fish and crustaceans; its use is not recommended

*Citric acid*    Antioxidant and acidifier; found in living organisms and in many natural foods; some reservations have been expressed about using it in conjunction with antibiotics

*Diphenyl (phenylbenzene)*    Mycostat used to prevent the formation of molds in crates of citrus fruit during shipping and storage; reservations have been expressed about its use chiefly because of the concrete possibility that relevant amounts of it may contaminate fruit juices

*Hexamethylenetetramine (HMT)*    Used as a preservative for fish, foods, and vinegar preserves; separates easily into formaldehyde and ammonia

*Phenylphenol, sodium phenylphenate*    Have the same uses as diphenyl with the same reservations

*Duodecyl gallate, propyl gallate*    Antioxidant agents widely used for the preservation of oils and fats

*Nordihydroguaiaretic acid*    Antioxidant for oils and fats

*Sodium* and *potassium nitrites, sodium* and *potassium nitrates*    Used to preserve freshness and color in meats and foods based on meat and fish; serious reservations have arisen over their use because there is evidence that they may cause the formation of carcinogenic nitrosamines in the organism

*Sodium, potassium, calcium propionates*    Fungicides used in the preservation of various foods

*Derivatives of sulfurous acid (sulfurous anhydride, sulfites, etc.)*    Antioxidant, antiseptic, and antibrowning agents widely used for the preservation of, among other things, wines and meats; some drawbacks due to the release of $SO_2$, which causes inflammation of the gastrointestinal tract

*Sorbic acid (E 200)*    Antimicrobial agent active at acid pH (pH 4.5)

*Salicylic acid*    Antimicrobial agent active at acid pH; widely used until a few years ago, has been recently banned as a food additive even in low concentrations; its ban is inconsistent with the large use, not only as a drug, of aspirin (acetylsalicylic acid)

*Tocopherols*    Extensively used as antioxidant agents in edible oils and fats

*Coloring Agents*

*Azorubin (E 122)*    Maximum daily dose (MDD) 0.5 mg/kg; research into its metabolism is recommended

*Amaranth (E 123)* As a result of studies conducted in the United States and the USSR has been recently banned in several countries

*Caramel (E 150)* Currently undergoing toxicological testing

*Quinoline yellow* MDD 0.5 mg/kg

*Beet red* Natural constituent of foods; the FAO is limiting its uses

*Brilliant black BN (E 151)* MDD 0–2.5 mg/kg; under toxicological examination

*Natural carotenes (E 160)* Under toxicological examination

*Chlorophylls* The MDD has been set at 15 mg/kg, pending the results of testing

*Thickening Agents*

*Lecithins* Commercial lecithins are a mixture of phospholipids prepared from vegetable oils, wheat, and other materials of animal origin; used as emulsions and stabilizers for bread, creams, margarine, and pastry; the normal daily diet furnishes 1–5 g of lecithins

*Agar-agar* A particular type of gelatin used as a thickener and stabilizer in, among other foods, whipped cream, jellies, and marmalades

*Alginates* Consist of sodium, ammonium, caclium, and potassium alginates; same uses as agar-agar; no reservations over human consumption of alginates in amounts lower than 50 mg/kg

*Methylcellulose* used as a thickener; no known drawbacks

*Sweeteners*

*Saccharin* Still the most common no-calorie sweetener; reservations have been expressed over its use as a result of studies that suggest that saccharin may cause bladder tumors

*Sorbitol* Used as a sweetener, stabilizer, and humectant in products for diabetics and in many dietetic foods; absorbed much more slowly than glucose and fructose

*Cyclamates* Very extensively used in the past, their use is currently prohibited

# Appendix B    Pesticides

The most important types of pesticides are insecticides, herbicides, fungicides, and rodenticides.

## Insecticides

The first substances to be used as insecticides were inorganic. Records show that arsenic pentoxide was already used against ants in 1681, and arsenic compounds have played an important part in the development of insecticides in spite of their toxicity to warm-blooded animals. They act through ingestion, and their agricultural use lasted practically until the introduction in 1945 of organic insecticides. Since then they have lost their preeminent role.

We can divide organic insecticides into natural and synthetic and establish a series of generations: The first generation consists of all the insecticides developed before DDT and the organophosphates; the second generation includes all subsequent insecticides; and the third consists of ecdysonics and juvenile hormones, both of which are very selective compounds and thus particularly suited to integrated programs of pest control. Insecticides of microbial origin, which, unlike ecdysonics and juvenile hormones, have been commercially developed, are also included in the third generation by some authors. For a general classification of organic insecticides see tables B.1 and B.2; table B.3 lists diseases caused or transmitted by insects.

## Herbicides (or Weed Killers)

Our century has seen a tremendous increase in the use of chemical herbicides, particularly after the introduction in 1945 of 2,4-D (2,4-dichlorophenoxyacetic acid). The most important steps in the development of herbicides are the following: introduction from 1896 to

**Table B.1**
Natural organic insecticides

| Origin | Insecticidal substance |
| --- | --- |
| Extracted from vegetables | Pyrethrins (from *Chrysanthemum cinerariaefolium*); nicotine (a tobacco alkaloid); quassin (from *Quassia amara* and *Picrasma excelsa*); rotenone compounds (from *Derris elliptica, Derris malacensis, Lonchocarpus utilis,* etc.); ryanodine (from *Ryania speciosa*) |
| Produced from bacteria, fungi, and animal organisms | The only industrial development comes from *Bacillus thuringiensis;* used as viable spores that infect many species of lepidopterans and as toxins that form by fermentation during production of the preparations; unlike the endotoxin, the exotoxin, which is thermostable, is also toxic to warm-blooded animals |
| Produced from oil and tar distillation | Phytotoxicity depends on the amount of unsaturated hydrocarbons and on the presence of impurities |

1900 of selective sprays for the control of weeds infesting cereal crops; discovery of the transfer of arsenic in *Convolvulus arvensis*, made by G. Gray in California in 1919; introduction of sodium chlorate as a weed killer to spread on the soil (1925); introduction of 2,4-D (1945).

One classification of herbicides divides them into selective and nonselective types. When applied to a mixed population of plants, the former destroy some species with little or no damage to the others, while the latter are totally destructive. Both types can be used either on leaves or for broadcasting on the soil. The herbicides that are applied to leaves are of two types: contact herbicides (or defoliants) and transferable (or systemic). The latter are absorbed by the leaves into vascular tissue and then transferred to other parts of the plant.

Systemic selective herbicides include 2,4-D and derivatives; 2,4,5-T (2,4,5-trichlophenoxyacetic acid); Atrazine (2-chloro-4-ethylamino-6-isopropylamino-1,3,5-triazine); Dalapon ($\alpha,\alpha$-dichloropropionic acid); Picloram (4-amino-3,5,6-trichloropicolinic acid).

Many selective herbicides are applied to the soil even before weeds start sprouting. They have multiplied in number since 1960 and include, to mention only a few, derivatives of benzoic acid, the chloroacetamides, the thiocarbamates, and phthalamic acids. Among nonselective contact herbicides we should mention sodium arsenate and chlorate, penta- and metaborates in solution, DN (4,6-dinitro-*o*-cresol), aromatic oils, PCP (pentachlorophenol), acrolein, diquat, paraquat, and erbon. Nonselective transferable herbicides include ar-

**Table B.2**
Synthetic organic insecticides

| Type | Insecticidal substances |
| --- | --- |
| Dinitrophenol derivatives | Dinitro-*o*-cresol (DNOC); dinocap; binapacryl; dinobuton |
| Thiocyanic and cyanic compounds | Lethane; Thanite; for household use along with the pyrethrins |
| Aliphatic chlorinated hydrocarbons | Fumigants for soil and goods; methyl bromide (which generally contains chloropicrin); carbon tetrachloride; trichloronitromethane (or chloropicrin); 1,2-dibromoethane; 1,2-dichloroethane; |
| Aromatic chlorinated hydrocarbons | DDT; Methoxychlor (DMDT); Rhothane (DDD or TDE); Perthane |
| Halogenated alicyclic hydrocarbons | γ-Hexachlorocyclohexane (Gammexane), Lindane, BHC, HCH; Toxaphene; chlordane (Octachlor); Heptachlor; Thiodan (Endosulfan); Aldrin; Dieldrin; Endrin, Mirex; Kepone (Chlordecone) |
| Phosphates, phosphonates, phosphoroamidates[a] | Dichlorvos (DDVP, Nogos, Nuvan, Vapona); Mevinphos (Phosdrin); Tetrachlorvinphos (Gardona); Chlorfenvinphos (Birlane); Phosphamidon (Dimecron); Monochrotophos (Azodrin, Nuvacron); Trichlorphon (Neguvon, Dipterex); Fenamiphos (Nemacur P); Phosfolan (Cyolane) |
| Phosphorothionates, phosphonothionates[a] | Parathion (E 605); Methylparathion; Fenitrothion (Sumithion); Bromophos (Nexion); Fenthion (Baytex, Lebaycid); Diazinone (Basudin); Chlorpyrifos (Dursban); Piriminphos-methyl (Actellic); Abate; Demeton-0; Phoxim (Valexon, Baythion); EPN; Surecide; Leptophos (Phosvel, Abar) |
| Phosphorothiolates, phosphoroamidothiolates[a] | Demeton-S-methyl (Metasystox i); Vamidothion (Kilval); Acaphate (Orthene); Methamidophos (Monitor, Tamaron) |
| Phosphorothiolothionates, phosphonothiolothionates[a] | Malathion (Cythion); Phenthoate (Cidial, Tanone, Elsan); Dimethoate (Rogor, Perfekthion); Phorate (Thimet); Disulfoton (Disyston); Azinphos-methyl (Gusathion M, Guthion); Methidathion (Supracid; Ultracid); Fonofos (Dyfonate) |
| Carbamates | Promecarb (Carbamult); Bux (Bux-ten); Aminocarb (Matacil); Propoxur (Baygon); Unden; Carbaryl (Sevin); Carbofuran (Furadan); Pirimicarb (Pirimor); Methomyl (Lannate); Aldicarb (Temik); Talcord |
| Organofluorine derivatives | Fumette; Bis (2-fluoroethoxy) methane; Nissol; Fluenetil |
| Various organic derivatives | Carbon disulfide; ethylene oxide (Cartox); naphthalene; phenothiazine (and similar compounds); 2-isovaleryl-1,3-indanedione (Valone); 2,pivalyl-1,3-indanedione (Pyvalin) |
| Synthetic pyrethrins | Allethrin; Cyclethrin; Barthrin; Tetramethrin; Permethrin (NRDC 143) |

a. Organophosphorous compounds.

senical acids, chlorates, thiocyanates, sulfamates, aminotriazole, TCA (trichloroacetic acid), and 2,3,6-TBA (2,3,6-trichlorobenzoic acid).

*Fungicides and Fungistats*

There is a distinction between fungicides and fungistats for human and animal use and those for agricultural use. Among fungicides for human and animal use, cycloheximide (Actidione), an antibiotic isolated from *Streptomyces griseus*, is active against mycelia and yeasts. It is toxic to humans, although it has proved effective against *Cryptococcus neoformans*, an organism that causes serious infections of the central nervous system. Undecylenic acid, sodium propionate, and creams based on sulfur and salicylic acid have proven effective against skin herpes; sodium propionate has shown some usefulness in the treatment of vulvovaginitis due to some types of *Candida*. These substances are of no use against fungal infections of the hair and nails or against skin infections caused by *Trichophyton rubrum*. The following drugs have proven effective in the treatment of human mycotic diseases: nystatin (from *Streptomyces noursei*), amphotericin B, and griseofulvin (from *Penicillium griseofulvum*).

Fungicides for agricultural use are both organic and inorganic. Inorganic fungicides are the Bordeaux mixture (a solution of copper sulfate and a suspension of calcium hydroxide) and sulfur. Organic fungicides have been gaining in importance since 1934. (The fungicidal properties of derivatives of dithiocarbamic acid were first observed in 1931.) Among them, we might mention Zineb, Ziram, and the tetraalkylthiuram disulfides. Other organic fungicides are captan, glyodin, chloranil, and dichlone to mention a few. Some antibiotics are also used as fungicides, notably streptomycin (against tobacco Peronospora) and Actidione (against leaf spot of the cherry). Some guanidine derivatives have also proved effective in this capacity. Table B.4 lists the most important diseases caused by fungal infections of marketed vegetables.

*Rodenticides*

Rodenticides are used to kill rodents as well as to control other mammals such as ground hogs, rabbits, and hares. *Anticoagulants* are used particularly for rats and mice and do not pose much threat to humans and domestic animals (e.g., Diphacin, Fumarin, PMP, Pival, Prolin, and Warfarin). *Poisons*, commonly used in baits, may cause death through only one ingestion. Among them are ANTU [1-(1-

**Table B.3**
Principal diseases caused or transmitted by insects

| Disease | Insect vector | Current methods of control | Promising methods of control | Geographical spread |
|---|---|---|---|---|
| Chagas's disease | Reduviid bug (*Rhodinus prolixus, Triatoma, Panstrongylus*) | Spraying with lindane or dieldrin | Improving housing conditions to eliminate breeding sites | Central and South America; Mexico |
| Dengue | Mosquito (*Aedes aegypti*) | Larvicides (Abate, malathion); destruction of habitat; control of adult mosquito | Genetic control | Middle East; tropical regions of Africa, Asia, Australia, and America; islands of the Pacific Ocean |
| St. Louis encephalitis | Mosquito (*Culex tarsalis, Culex quinquefasciatus*, etc.) | Urban sanitation measures (destruction of breeding sites); insecticides for the control of larva and adult mosquito | Water treatments; introduction of predacious fish | North and South America |
| Bancroftian filariasis (urban) | Mosquito (*Culex fatigans* and others) | Urban sanitation measures; chlorinated hydrocarbons; fenthion | Permanent sanitation measures; genetic and biological control; natural toxins | Tropical Africa; Eastern Mediterranean; South America; South-East Asia; Western Pacific |
| Leishmaniasis | Sandfly (*Phlebotomus*) | Spraying with chlorinated hydrocarbons | | India; China; Middle East; Africa; Latin America |
| Malaria | Mosquito (*Anopheles*) | Spraying with DDT; control of the larva | Predacious fish (*Gambusia*, etc.); alternative insecticides (propoxur, fenitrothion); environmental sanitation | Most tropical areas |
| Onchocerciasis | Black fly (*Simulium*) | Control of the larva (DDT, metoxychlor); aerial spraying | | West Africa; Latin America |
| Bubonic plague | Flea | Insecticidal dusts; rat control | Sanitation measures and education in personal hygiene | Foci in America, Asia, and Africa |
| Schistosomiasis | Water snail | Water treatments; molluscicides | Introduction of competitive species and predacious fish | Africa; Brazil; Puerto Rico; Venezuela; Asia |

| Rickettsial diseases | Arthropod (body lice, fleas, mites, ticks) | Repellents; dusting of the infested area | Vaccine; antibiotics; removal of insects from body | Americas; Asia; Africa |
|---|---|---|---|---|
| Trypanosomiasis | Tsetse fly (*Glossina palpalis, Glossina morsitans*, etc.) | Application of residual insecticides to the vegetation (DDT, lindane) | Sterilization of males; alternative insecticides | Tropical Africa |
| Typhus | Body louse | Application of dusting powders (DDT, lindane) to body and clothing | Improved personal hygiene; sanitation measures; vaccine; antibiotics | Eastern Europe; Mediterranean basin; South and North Africa; South America |
| Yellow fever | Mosquito (*Aedes*) | Vaccination; larvicides (Abate, malathion); destruction of habitat; control of adult mosquito | Genetic control | Tropical America; Western and Central Asia |

Source: *WHO Chronicle* 25:5 (1971).

**Table B.4**
Major plants in world commerce and their diseases

| Plant | Disease |
| --- | --- |
| **Cereals** | |
| Rice | Blast, bacterial blight |
| Wheat | Rusts, smuts |
| Maize | Stem, root rots |
| Sorghum | Smut |
| Barley | Helminosporial leaf spot, root rots |
| **Sugar crops** | |
| Sugar cane | Virus |
| Sugar beet | Virus, cercosporal leaf spot |
| **Root crops** | |
| Potato | Late blight, viruses |
| Sweet potato | Stem rot |
| Cassava | Mosaic |
| **Legumes** | |
| Common bean | Viruses, bacterial blights, root rots |
| Soy bean | Root rot |
| Peanut | Leaf spots, root rot |
| **Tree crops** | |
| Coconut | Practically none |
| Banana | Wilt, sigotoka |
| **Nonfood** | |
| Rubber | Leaf blight |
| Coffee | Rust |
| Cotton | Wilt, and rots of seedlings and bolls |
| Tobacco | Blue mold, viruses |
| Tea | Blister blight |

Source: J. G. Horstall, "Fungicides: Past, present, and future," *Chemtec* 7(5):302 (1977).

naphthyl)-2-thiourea], red squill (obtained from the bulb of the liliacea *Urginea maritima*; rats keep it down because they are incapable of vomiting), fluoroacetamide, strychnine alkaloids, strychnine sulfate, thallium sulfate, sodium fluoroacetate, and zinc phosphide. All of these products lack valid antidotes. Some selective rodenticides are more useful than others for controlling certain species. Norbormide, for example, is toxic to the genus *Rattus* but not to others, even in large doses. The most commonly used *gases* as fumigants are calcium cyanide, carbon monoxide, carbon disulfide, chloropicrin, sulfur dioxide, and methyl bromide. Among the *dusts* are ANTU, sodium fluorosilicate, DDT, Warfarin, dieldrin, and endrin.

# Appendix C    Nutrients Necessary to Vegetable Growth

To grow, plants need (at least) 16 nutritive elements, which they obtain from the air, water, and soil as well as from organic and inorganic fertilizers. Plants consist mostly of carbon, oxygen, and hydrogen, three elements that they obtain from water and air. Carbon comes from the carbon dioxide ($CO_2$) found in the air (approximately 0.04%). The other thirteen elements, which can be obtained either from the soil or from fertilizers added to it, may be divided in three groups: principal elements, secondary elements, and trace elements. The *principal elements* are nitrogen, phosphorus, and potassium.

Nitrogen, which constitutes 1–4% of dried vegetable matter, is absorbed both as nitrate and in ammoniacal form and converted to amino acids and proteins. Phosphorus, which is generally measured as phosphorus pentoxide ($P_2O_5$), constitutes 0.1–0.4% of dried vegetable matter. It is essential to cell division and to tissue development. Potassium (generally measured as $K_2O$) constitutes 0.5–4% of dried vegetable matter. It appears to have a fundamental role in the utilization of the other elements. *Secondary elements* are calcium, magnesium, and sulfur, which are found in smaller but still significant amounts. *Trace elements* are found in very low levels and form ligands with key molecules in the most important biochemical processes.

# Appendix D    Weapons

Generally speaking, modern weapons are based on the explosive properties of some substances. In this sense, all weapons are chemical, but in the context of this book, by chemical weapons we mean the toxic compounds that have been used, or may be expected to be used, directly against living beings.

Not much is known about chemical weapons since they fall into the category of military secrets. As a consequence, the picture we present is necessarily *incomplete* and refers almost exclusively to the western world.

Chemical weapons may be classified as follows: lethal agents; disabling agents; neurotoxic agents; and defoliants and soil sterilizers. Tables D.1–D.4 summarize all outstanding information about the four types of chemical weapons. Table D.5 lists the characteristics of selected lethal and disabling agents.

Military research budgets are presented in tables D.6 and D.7 and in figures D.1–D.3.

**Table D.1**
Lethal chemicals used in warfare and stockpiled for military use

| Chemical name | Common name | Effective dose[a] | Remarks[b] |
|---|---|---|---|
| **Asphyxiating agents (with delayed effect)** | | | |
| Trichloronitromethane | Chloropicrin, Klop | 20,000 | AA, E |
| Chlorine | | 20,000 | A |
| Bromine | | | A |
| Perchloromethylmercaptan | | 15,000 | A |
| Methylchlorosulfonate | | 15,000 | A |
| Ethylchlorosulfonate | | 10,000 | A |

**Table D.1** (Continued)

| Chemical name | Common name | Effective dose[a] | Remarks[b] |
|---|---|---|---|
| Phenylcarbilamine | | 10,000 | A |
| Cyanoformic esters | Zyklon | | A |
| Chlormethyl chloroformiate | C-Stoff | 10,000 | A |
| Dichloromethyl ether | | 10,000 | A |
| Dibromomethyl ether | | 10,000 | A |
| Phenyldibromoarsine | | 5,000 | A |
| Carbonyl chloride | Phosgene | 3,200 | AA, E, F |
| Trichloromethyl | | | |
| Chloroformiate | Diphosgene | 3,200 | AA |
| **Vesicant agents (with delayed effect)** | | | |
| Dimethyl sulfate | D-Stoff | | A |
| Phenyldichloroarsine | PD | 2,000 | A, E |
| Ethyldichloroarsine | Dick | 1,500 | A |
| Ethyldibromoarsine | | | A |
| 2-chlorovinyldichloroarsine | Lewisite | 1,500 | D, E |
| Ethyldi(2-chloroethyl)amine | HN 1 | 200 | E |
| Methyldi(2-chloroethyl)amine | HN 2 | 200 | E |
| Tri(2-chloroethyl)amine | HN 3 | 200 | E |
| Di(2-chloroethyl)sulfide | Yperite, mustard gas | 200 | AA, C, D, E, F, H |
| Di(2-chloroethylthioethyl)sulfide | T | 100 | E, H? |
| 1,2-di(2-chloroethylthio)ethane | Sesquiyperite | 50 | E |

Source: J. P. Perry Robinson, "Gli aggressivi chimici," *Scienza e tecnica* 70:397 (1970).
a. Evaluation of the effective dose (in mg-min/m³) takes into account the exposure time on the battlefield necessary to cause damage to humans (death in the case of asphyxiating gases, gases toxic to the blood, and nerve gas; blisters and blindness in the case of vesicants; immobilization in the case of disabling agents). The figures given are averages of data taken from many different sources. An effective dose of 100 mg-min/m³ corresponds to an exposure time of 10 minutes at a concentration of 10 mg/m³ of the chemical and to an exposure time of 1 minute at a concentration of 100 mg/m³.
b. The letters denote the following: A, used in World War I; AA, successfully used in World War I; B, used in the Russian Civil War, 1917–1921; C, used in the Italo-Ethiopian war, 1935–1936; D, used in the Sino-Japanese war 1937–1944; E, stockpiled during World War II; F, used in Yemen's civil war in 1966 and 1967; G, used in the Viet Nam War, from 1964; H, currently stockpiled; I, used in Northern Ireland in 1969 to quell riots.

**Table D.2**
Disabling agents used in warfare and stockpiled for military use

| Chemical name | Common name | Effective dose[a] | Remarks[b] |
|---|---|---|---|
| *o*-dianisidine chlorosulfonate | Ni-Stoff | | A |
| Chloracetone | | 100 | A |
| Acrolein | Papite | 100 | A |
| Xylene bromide | T-Stoff | 50 | AA |
| Benzyl bromide | | | |
| Ethyl bromoacetate | | 50 | A |
| Ethyl iodoacetate | SK | 50 | AA |
| Benzyl iodide | | 50 | A |
| Dichloroformossima | Phosgenossima | 25 | E? |
| Methylbromoethylketone | | 20 | A |
| Iodoacetone | | 20 | A |
| Bromoacetone | B-Stoff, BA | 10 | AA |
| ω-chloroacetophenone | CN | 10 | E, G, H |
| α-bromobenzyl cyanide | CA | 5 | A |
| Diphenylchlorarsine | Clark I, DA | 5 | A, D, E |
| Diphenylcyanoarsine | Clark II, DC | 5 | A, D, E |
| Diphenylaminechlorarsine | DM, adamsite | 2 | B, E, G, H |
| *o*-chlorobenzalmalononitrile | CS | 2 | G, H, I |

Source: J. P. Perry Robinson, "Gli aggressivi chimici," *Scienza e tecnica* 70:397 (1970).
a. In mg-min/m³.
b. For an explanation of the letters see table D.1.

**Table D.3**
Neurotoxic agents (nerve gas) used in warfare and stockpiled for military use

| Chemical name | Common name | Effective dose[a] | Remarks[b] |
|---|---|---|---|
| Ethyl NN-dimethylaminocyanophosphate | Tabun | 200 | E |
| Isopropyl methylfluorophosphonate | Sarin | 100 | H |
| 1,2,2-trimethylpropyl methylfluoro-phosphonate | Soman | 50 | H? |
| Alkyl S-dialkylaminoethylmethyl-phosphonothiolates | Agents V | 10 | H |

Source: J. P. Perry Robinson, "Gli aggressivi chimici," *Scienza e tecnica* 70:397 (1970).
a. in mg-min/m³.
b. For an explanation of the letters see table D.1.

**Table D.4**
Defoliants and soil sterilizers

| Chemical term | Common name | Remarks |
|---|---|---|
| 2,4-dichlorophenoxyacetic acid[a] | 2,4-D | $LD_{50}$ by mouth for mice = 375 mg/kg; discovered in 1942 by E. J. Kraus; used in Viet Nam |
| 2,4,5-trichlorophenoxyacetic acid[a] | 2,4,5-T | A plant hormone with herbicidal properties; similar in toxicity to 2,4-D; used in Viet Nam |
| 4-amino-3,5,6-trichloropicolinic acid[a] | Picloram | $LD_{50}$ in female mouse = 2,000–4,000 mg/kg |
| sodium cacodylate[a] | | Used in Viet Nam as a defoliant from 1962 |
| 2,4-dinitrophenol[a] | 2,4-DNP | Highly toxic to humans |
| 4,6-dinitro-2-methylphenol[b] | DNOC | Acts as a selective herbicide |
| 3-(*p*-chlorophenyl)-1,1-dimethylurea[b] | Monuron CMU | Acts on young plants |
| 5-bromo-3-butyl-6-methyluracil[b] | Bromacile | |

Source: Nguyên Dang Tam, "La guerre chimique," *La Recherche* 1:442 (1970).
a. Defoliant.
b. Soil sterilizer.

**Table D.5**
Characteristics of selected lethal or disabling chemical agents

| Characteristics[a] | VX[b] | Tabun | Phosgene | Yperite | CS | BZ[c] | Botulinial toxin A | Staphylococca enterotoxin |
|---|---|---|---|---|---|---|---|---|
| Classification | Nerve gas (lethal) | Nerve gas (lethal) | Asphyxiating (lethal) | Vesicant (lethal) | Irritant (disabling) | Psychotropic (disabling) | Toxin (lethal) | Toxin (disabling) |
| Physical state (at 20°C) | Liquid | Liquid | Gas | Liquid | Solid | Solid | Solid | Solid |
| Mode of dissemination | Liquid or aerosol | Liquid, gas, or aerosol | Gas | Liquid or gas | Aerosol | Aerosol | Aerosol | Aerosol |
| Onset of initial symptoms | Very fast | Very fast | 4 hours | 4 to 6 hours | Instantaneous | 4 to 12 hours | 12 hours | Very fast |
| Physiopathologic mechanism | Meiosis, bronchial constriction, cholinesterase poisoning | Similar to VX | Tearing, bronchial constriction, pulmonary edema, vesication, coma, convulsions | Acute conjunctivitis, blindness, skin lesions, neuro muscular paralysis | Eye and skin irritation, chest pains, dyspnea. Effects are temporary | Blurred vision, dazedness, mental confusion. Effects are temporary | Nausea, vertigo, vomiting, paralysis, secretionary disorders | Vomiting, diarrhea, fever, anticholinergic poisoning |
| First aid | Atropine, oxime, artificial respiration | Atropine, artificial respiration | Oxygen, stimulants | Ointments, stimulants | Fresh air | Physostigmine | Antitoxin | Antitoxin and anatoxin |
| Protective measures | Mask and protective clothes | Mask and protective clothes | Mask | Mask, protective clothes, and ointments | Mask | | | |
| Decontamination | Calcium chloride, diluted alkali solutions, hot soapy water | See VX | Aeration of closed spaces | Lime chloride | None | | | |

Source: J. P. Perry Robinson, "Gli aggressivi chimici," *Scienza e tecnica* 70:397 (1970).
a. For chemical terms see Table D.1.
b. The formula of VX is a military secret.
c. The formula of BZ is a military secret; it is believed to be a phenylglycolic ester of 3-quinoquidinole.

**Table D.6**

World military expenditures (in millions of dollars)[a]

|  | 1956 | 1957 | 1958 | 1959 | 1960 | 1961 | 1962 | 1963 | 1964 | 196 |
|---|---|---|---|---|---|---|---|---|---|---|
| USA | 68,234 | 69,584 | 69,622 | 70,004 | 68,130 | 70,937 | 76,943 | 75,824 | 73,326 | 7 |
| Other NATO | 29,245 | 29,817 | 27,301 | 29,830 | 31,050 | 32,241 | 35,397 | 36,697 | 37,241 | 3 |
| **Total NATO** | **97,479** | **99,401** | **96,923** | **99,834** | **99,180** | **103,178** | **112,340** | **112,521** | **110,567** | **110** |
| USSR | 31,600 | 31,300 | 30,500 | 36,000 | 32,700 | 40,800 | 44,600 | 48,900 | 46,700 | 44 |
| Other WTO[b] | 2,600 | 2,700 | 2,900 | 3,000 | 2,958 | 3,250 | 4,147 | 4,469 | 4,471 |  |
| **Total WTO** | **34,200** | **34,000** | **33,400** | **33,000** | **35,658** | **44,050** | **48,747** | **53,369** | **51,171** | **4** |
| Other Europe | 2,880 | 3,160 | 3,225 | 3,300 | 3,300 | 3,546 | 3,867 | 3,999 | 4,226 |  |
| Middle East | 975 | 1,025 | 1,225 | 1,325 | 1,340 | 1,440 | 1,600 | 1,785 | 2,065 |  |
| South Asia | 975 | 1,100 | 1,100 | 1,075 | 1,090 | 1,150 | 1,494 | 2,317 | 2,287 |  |
| Far East (excl China) | 2,725 | 2,900 | 3,100 | 3,275 | 3,375 | 3,525 | 3,740 | 3,926 | 4,249 |  |
| China | [9,100] | [9,800] | [9,100] | [10,100] | [10,100] | [11,800] | [13,700] | [15,500] | [18,400] | [19 |
| Oceania | 1,058 | 974 | 976 | 1,024 | 1,018 | 1,006 | 1,039 | 1,166 | 1,356 |  |
| Africa (excl Egypt) | 260 | 300 | 275 | 325 | 390 | 575 | 860 | 961 | 1,149 |  |
| Central America | 300 | 350 | 375 | 400 | 435 | 459 | 509 | 545 | 580 |  |
| South America | 2,340 | 2,515 | 2,600 | 2,135 | 2,200 | 2,139 | 2,168 | 2,256 | 2,204 |  |
| **World total** | **152,292** | **155,525** | **152,299** | **155,793** | **158,086** | **172,868** | **190,064** | **198,345** | **198,254** | **198** |

Source: SIPRI, *World Armaments and Disarmament,* Stockholm (1977).
a. Expenditures are given at fixed 1973 prices and at 1973 exchange rates. Only the last column is at current prices.
b. At current prices and on the basis of the Benoit-Lubell exchange rates.

**Table D.7**

NATO military expenditures (in millions of dollars)[a]

|  | 1956 | 1957 | 1958 | 1959 | 1960 | 1961 | 1962 | 1963 | 1964 | 1965 |
|---|---|---|---|---|---|---|---|---|---|---|
| **North America** |  |  |  |  |  |  |  |  |  |  |
| Canada | 3,099 | 2,903 | 2,703 | 2,524 | 2,512 | 2,584 | 2,689 | 2,502 | 2,604 | 2,325 |
| USA | 68,234 | 69,584 | 69,622 | 70,004 | 68,130 | 70,937 | 76,943 | 75,824 | 73,326 | 72,928 |
| **Europe** |  |  |  |  |  |  |  |  |  |  |
| Belgium | 771 | 806 | 799 | 807 | 824 | 834 | 888 | 920 | 981 | 957 |
| Denmark | 360 | 379 | 370 | 361 | 404 | 411 | 502 | 508 | 524 | 556 |
| France | 7,639 | 7,929 | 7,321 | 7,469 | 7,699 | 7,935 | 8,229 | 8,087 | 8,311 | 8,446 |
| FR Germany | 4,600 | 5,566 | 4,141 | 6,611 | 7,148 | 7,535 | 9,562 | 10,749 | 10,301 | 10,180 |
| Greece | 281 | 247 | 242 | 251 | 266 | 258 | 262 | 268 | 279 | 302 |
| Italy | 1,924 | 1,991 | 2,033 | 2,121 | 2,204 | 2,279 | 2,500 | 2,787 | 2,853 | 2,961 |
| Luxembourg | 16 | 17 | 17 | 16 | 10 | 11 | 14 | 13 | 17 | 17 |
| Netherlands | 1,448 | 1,352 | 1,190 | 1,060 | 1,168 | 1,360 | 1,447 | 1,466 | 1,595 | 1,554 |
| Norway | 352 | 373 | 348 | 368 | 350 | 381 | 421 | 438 | 444 | 515 |
| Portugal | 217 | 223 | 229 | 257 | 266 | 427 | 485 | 474 | 517 | 517 |
| Turkey | 386 | 375 | 387 | 445 | 469 | 506 | 532 | 541 | 585 | 621 |
| UK | 8,152 | 7,656 | 7,521 | 7,530 | 7,730 | 7,720 | 7,886 | 7,944 | 8,230 | 8,206 |
| **Total NATO** | **97,479** | **99,401** | **96,923** | **99,834** | **99,180** | **103,178** | **112,340** | **112,521** | **110,567** | **110,085** |
| **Total NATO** (excl. USA) | **29,245** | **29,817** | **27,301** | **29,830** | **31,050** | **32,241** | **35,397** | **36,697** | **37,241** | **37,157** |
| **Total NATO Europe** | **26,146** | **26,914** | **24,598** | **27,306** | **28,538** | **29,657** | **32,708** | **34,195** | **34,637** | **34,832** |

Source: SIPRI, *World Armaments and Disarmament,* Stockholm (1977).
a. Expenditures are at fixed 1973 prices and at 1973 exchange rates. The last column is at current prices.

| 6 | 1967 | 1968 | 1969 | 1970 | 1971 | 1972 | 1973 | 1974 | 1975 | 1976 | 1975X |
|---|---|---|---|---|---|---|---|---|---|---|---|
| ,993 | 100,363 | 103,077 | 98,698 | 89,065 | 82,111 | 82,469 | 78,358 | 77,383 | 75,068 | 77,373 | *90,948* |
| ,325 | 38,980 | 37,806 | 37,638 | 38,385 | 40,412 | 42,619 | 43,326 | 44,543 | 45,651 | 46,859 | *58,194* |
| ,318 | 139,343 | 140,883 | 136,336 | 127,450 | 122,523 | 125,088 | 121,684 | 121,926 | 120,719 | 124,232 | *149,142* |
| ,000 | 50,800 | 58,600 | 62,200 | 63,000 | 63,000 | 63,000 | 63,000 | 61,900 | 61,100 | 61,100 | *61,000* |
| ,833 | 5,252 | 6,387 | 7,012 | 7,498 | 7,974 | 8,240 | 8,808 | 9,444 | 10,207 | 11,007 | *10,207* |
| ,833 | 56,052 | 64,987 | 69,212 | 70,498 | 70,974 | 71,240 | 71,808 | 71,344 | 71,307 | 72,107 | *71,307* |
| ,422 | 4,420 | 4,560 | 4,740 | 4,864 | 4,983 | 5,288 | 5,382 | 5,650 | 5,658 | 5,900 | *7,761* |
| ,830 | 3,700 | 4,450 | 5,140 | 6,175 | 6,425 | 8,820 | 11,468 | 15,737 | 19,875 | 21,835 | *25,164* |
| ,313 | 2,101 | 2,176 | 2,312 | 2,403 | 2,856 | 3,100 | 2,775 | 2,611 | 2,804 | 3,210 | *3,638* |
| ,862 | 5,348 | 5,949 | 6,387 | 6,917 | 7,589 | 8,005 | 8,032 | 8,200 | 8,250 | 8,700 | *10,855* |
| ,800] | [21,800] | [22,800] | [24,600] | [27,200] | [28,200] | [27,300] | [27,300] | [27,300] | [27,300] | [27,300] | *[32,300]* |
| ,779 | 1,937 | 2,101 | 2,129 | 2,125 | 2,125 | 2,131 | 2,102 | 2,177 | 2,157 | 2,097 | *2,597* |
| ,382 | 1,712 | 1,984 | 2,376 | 2,514 | 2,776 | 2,869 | 3,096 | 3,676 | 4,550 | 5,200 | *6,039* |
| 614 | 659 | 738 | 718 | 754 | 771 | 791 | 812 | 811 | 900 | 950 | *1,105* |
| ,687 | 3,170 | 3,006 | 3,149 | 3,230 | 3,700 | 3,781 | 4,003 | 3,550 | 4,700 | 4,500 | *4,417* |
| ,840 | 240,242 | 253,634 | 257,099 | 254,130 | 252,922 | 258,413 | 258,462 | 262,982 | 268,220 | 276,031 | *314,325* |

| 66 | 1967 | 1968 | 1969 | 1970 | 1971 | 1972 | 1973 | 1974 | 1975 | 1976 | 1975X |
|---|---|---|---|---|---|---|---|---|---|---|---|
| 2,386 | 2,562 | 2,415 | 2,276 | 2,392 | 2,403 | 2,409 | 2,408 | 2,582 | 2,546 | 2,723 | *3,074* |
| 6,993 | 100,363 | 103,077 | 98,698 | 89,065 | 82,111 | 82,469 | 78,358 | 77,383 | 75,068 | 77,373 | *90,948* |
| 977 | 1,019 | 1,066 | 1,067 | 1,136 | 1,152 | 1,215 | 1,259 | 1,311 | 1,417 | 1,480 | *1,854* |
| 548 | 547 | 584 | 574 | 563 | 617 | 613 | 583 | 638 | 693 | 722 | *914* |
| 8,688 | 9,155 | 9,164 | 8,738 | 8,835 | 8,947 | 9,173 | 9,513 | 9,437 | 9,903 | 10,379 | *13,034* |
| 9,869 | 10,264 | 9,112 | 9,992 | 10,108 | 10,823 | 11,576 | 12,027 | 12,558 | 12,496 | 12,312 | *15,198* |
| 327 | 422 | 492 | 557 | 603 | 638 | 680 | 679 | 650 | 1,043 | (1,211) | *1,360* |
| 3,204 | 3,128 | 3,187 | 3,124 | 3,293 | 3,726 | 4,114 | 4,107 | 4,110 | 3,825 | 3,735 | *4,744* |
| 17 | 14 | 12 | 12 | 13 | 13 | 14 | 15 | 17 | 18 | 17 | *22* |
| 1,515 | 1,677 | 1,659 | 1,732 | 1,788 | 1,871 | 1,933 | 1,967 | 2,053 | 2,158 | 2,120 | *2,851* |
| 512 | 528 | 559 | 590 | 592 | 607 | 606 | 611 | 627 | 681 | 677 | *908* |
| 545 | 669 | 705 | 653 | 714 | 747 | 737 | 681 | 816 | 561 | 433 | *774* |
| 603 | 608 | 643 | 631 | 675 | 790 | 821 | 862 | 943 | (1,516) | (1,908) | *2,113* |
| 8,134 | 8,387 | 8,208 | 7,692 | 7,673 | 8,078 | 8,728 | 8,614 | 8,801 | 8,794 | 9,142 | *11,348* |
| 4,318 | 139,343 | 140,883 | 136,336 | 127,450 | 122,523 | 125,088 | 121,684 | 121,926 | 120,719 | 124,232 | *149,142* |
| 7,325 | 38,980 | 37,806 | 37,638 | 38,385 | 40,412 | 42,619 | 43,326 | 44,543 | 45,651 | 46,859 | *58,194* |
| 4,939 | 36,418 | 35,391 | 35,362 | 35,993 | 38,009 | 40,210 | 40,918 | 41,961 | 43,105 | 44,136 | *55,120* |

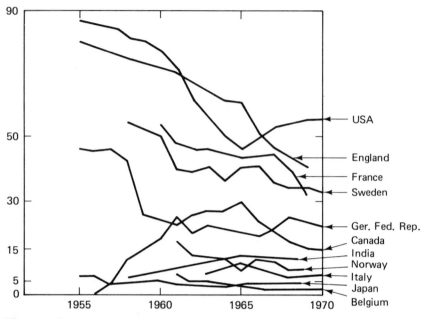

**Figure D.1**
The trend in the fraction of national research budgets devoted to military research and development 1955–1970. [From P. Vautier, "La recherche militaire: un premier bilan mondial," *La Recherche* 4:388 (1973)]

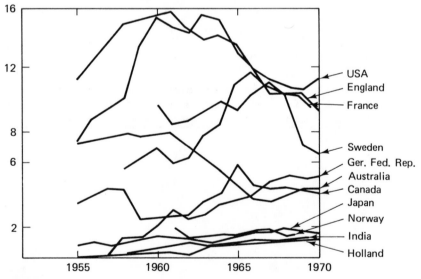

**Figure D.2**
The trend in the fraction of national defense budgets devoted to military research and development 1955–1970. [From P. Vautier, "La recherche militaire: un premier bilan mondial," *La Recherche* 4:388 (1973)]

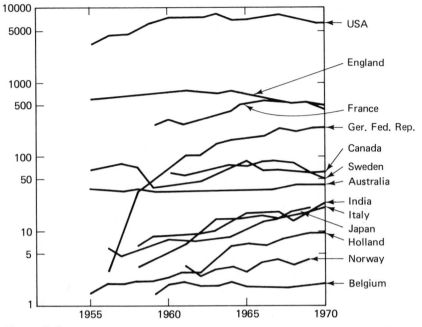

**Figure D.3**
The trend in military research and development expenditures 1955–1970 in
millions of dollars (1963 value) on a logarithmic scale. [From P. Vautier, "La
recherche militaire; un premier bilan mondial," *La Recherche* 4:388 (1973)]

# Glossary

**Additive**
A substance added to another to impart or improve desirable properties or suppress undesirable properties

**Aerosol**
A suspension of fine solid or liquid particles in gas; clouds and fog are suspensions of water or ice particles in the air; exhaust fumes are suspensions of solid particles

**Ames test**
A test performed on microorganisms *in vitro* to determine the mutagenic properties of a chemical

**Anabolic agents**
Drugs whose main clinical characteristic is the ability to facilitate the biosynthesis of proteins

**Analgesics**
Commonly called painkillers; drugs that exercise a particular action on the central nervous system whereby sensations of pain are relieved or suppressed

**Angiosarcoma**
Also called hemangiosarcoma; malignant tumor due to the proliferation of endothelial cells of blood vessels; in the benign form it is called hemangioendothelioma

**Angiotensins**
Decapeptide hormones that regulate the constriction of blood vessels and the secretion of aldosterone by the adrenal cortex

**Antibiotics**
Substances produced by certain microorganisms and able to inhibit (bacteriostats) or kill (bactericides) other microorganisms; the most

important antibiotics for clinical use are the penicillins, cephalosporins, tetracyclines, chloramphenicol, and the rifamycins

## Antihistamines
Drugs able to combat the effects of histamine and used for treating cold symptoms and allergic reactions caused by insect stings, foods, serums, and other drugs

## Antioxidants
Substances, generally organic, that slow down or inhibit reactions promoted by oxygen or peroxides; they may be found naturally in some products, but are usually added in the course of food processing

## Antivitamins
Or vitamin inhibitors, are substances that make vitamins ineffective by inhibiting their normal metabolic process

## Bacteria
Often classified as plants, these microorganisms should be considered apart from either the vegetable or animal realm; they are found everywhere (air, water, and soil) and many are pathogens

## Biomethylation
Process in living organisms whereby inorganic ions of heavy metals are turned into organic ions with the methyl group $CH_3$ bound to the metal in a covalent bond (e.g., methylmercury bromide, $CH_3HgBr$)

## Botulism
Food poisoning due to the ingestion of botulin, a toxin produced by *Clostridium botulinum* and *Clostridium parabotulinum,* two anaerobic microorganisms that develop in meats and preserved foods

## Carbamates
Salts or esters of carbamic acid, which cannot be isolated in the free state; esters are called urethanes; some derivatives have insecticidal and fungicidal properties

## Chelating agents
Organic compounds, such as ethylenediamine, ethylenediaminetetracetic acid (EDTA), and dimethylglyoxime, containing atoms that form two or more coordinate bonds with metal in solution

## Chromatography
A process of separating a mixture of compounds by allowing a solution to seep through an adsorbent, so that each component becomes adsorbed in a different, often colored, layer

**Edema**

An abnormal accumulation of serous fluid in connective tissue or in body cavities; in the latter case it is also called dropsy

**EEC (European Economic Community)**

An organization founded with the Treaty of Rome (March 1957) by six Western European countries (Belgium, France, German Federal Republic, Italy, Luxemburg, and Netherlands) for the purpose of achieving the economic integration of the member countries through free trade, free movement of work force and capitals, and progressive integration of social and fiscal policies (three objectives now all but realized); since 1970 the EEC has been pursuing the goal of monetary, economic, and political unification; since 1973 Denmark, Great Britain, and Ireland have joined the Community—Austria, Greece, Iceland, Malta, Portugal, Sweden, Switzerland, Turkey and various African countries have associated themselves with it

**Embryogenesis**

The formation and development of the embryo

**Enzymes**

Complex proteins that act as biological catalyzers; they are produced by living cells and mediate or promote specific biochemical processes without being altered or destroyed

**EPA**

Environmental Protection Agency, the US federal agency responsible for all the problems associated with the pollution of the environment

**Estrogens**

Substances (such as sex hormones) that promote estrus and stimulate the development of primary and secondary sex characteristics in the female; there are natural and synthetic estrogens able to perform the biological activity of native estrogenic hormones

**Eutrophication**

Phenomenon of chronic pollution most prevalent in lakes, marshes and enclosed seas, and due to a large influx of plant nutrients (nitrates and phosphates), which cause an abnormal growth of plant life; the decomposition of dead organisms results in the depletion of oxygen and the formation of methane, ammonia, phosphine, and hydrogen sulfide, with a general degradation of water quality

**FAO**

Food and Agriculture Organization of the United Nations, established in 1945 with the support of 42 countries (member countries currently number over 100); the general objective is to promote the production and distribution of agricultural goods and foodstuffs, thereby contributing to the solution of the problem of malnutrition in large areas of the world; its activities include technical assistance, food research and nutritional planning on a world scale

**Favism**
Acute anemia characteristic of the Mediterranean region, which develops when a person with a deficiency of the enzyme glucose-6-phosphate dehydrogenase eats fava beans or inhales the pollen of *Vicia faba*

**FDA**
Food and Drug Administration, the US federal agency responsible for the safety of foods and drugs

**Fluorocarbons**
Hydrocarbons in which part or all of the hydrogen atoms are replaced with fluorine atoms; they may be liquid or gaseous and are used as refrigerants, lubricants, and propellants in aerosol cans

**Fly ash**
Small, solid particles of ash and soot generated by combustion processes and carried aloft by air currents

**Food chain**
Series of organisms in which each group constitutes the food of the next one; man is the last stage of any food chain; microorganisms, plankton, and vegetables are the first stages, followed by herbivores and carnivores

**Formulation**
A special mixture of basic chemicals and additives designed to obtain a given product

**Freeze-drying**
A process of quick freezing and quick drying under high vacuum used especially for food preservation

**Fungicides**
Pesticides based on a variety of chemicals that destroy molds or inhibit their growth on cultures, stored goods, wooden materials, and textiles

**Galenicals**
Vegetable preparations (such as infusions and extracts), as opposed to purified chemical drugs; named after Galen, who was the first to describe methods of isolating curative substances of natural origin

**Genetic mutation**
Modification of the genetic code that alters the sequence of nucleotides during DNA duplication, so that the new DNA molecule is no longer an exact copy of the original; the modified molecule may cause the synthesis of a different type of protein in the cell; temperature, some chemicals, and radiation are among the agents that can cause an increase in the incidence of genetic mutations

**GNP**
Gross national product, or the total value of the goods and services produced in a nation during a specific period (generally a year), including amortization of capital investments; the net national product is obtained by deducting this amortization

**Green Revolution**
Popular term for the greatly increased yield in the developing countries resulting from the introduction of selected varieties of grains (wheat, rice, maize, millet), the large use of fertilizers and pesticides, and soil irrigation

**Herbicides**
Commonly called weed killers, chemicals used to destroy or inhibit plant growth, particularly weeds; they may be broad spectrum or highly specific; hormonal herbicides are not directly toxic to weeds, but kill them by altering their metabolic balance

**Hormones**
Substances secreted by endocrine glands that circulate in body fluids and produce specific effects on the activity of cells remote from their point of origin

**Hydrometallurgy**
The chemical treatment of poor, rich, or artificially enriched ores by wet processes (as leaching)

**Hypoglycemics**
Drugs prescribed for the treatment of diabetes mellitus, a disorder of carbohydrate metabolism characterized by inadequate secretion of insulin and excessive amounts of sugar in the blood; besides insulin, which is administered intravenously, there are hypoglycemic compounds that are administered orally, such as sulfonylurea and biguanides

**Insecticides**
Natural or synthetic chemicals used to combat insects harmful to hu-

mans, domestic animals, materials (wood, leather, textiles, etc.), plants, agricultural products, and stored goods

## Ischemia
Localized tissue anemia due to obstruction of the blood flow or vaso-constriction

## Keratitis
Inflammation of the cornea characterized by the sensation of a foreign body in the eye, acute pain, photophobia, and tearing, distinguished in suppurative and nonsuppurative forms (the former of syphilitic or viral origin, the latter of genetic, congenital or tubercular origin); the final outcome in both cases is opacity of the cornea due to scarring

## Lag phase
The time elapsing between the acquisition of a pathogenic agent and the onset of disease

## Lathyrism
A diseased condition of man and domestic animals that results from poisoning by a substance found in some legumes (genus *Lathyrus*, especially *L. sativus*) and is characterized by spastic paralysis of the lower or hind limbs

## $LD_{50}$
Average lethal dose, that is, the dose expected to kill 50% of a test group of organisms

## Mass spectroscopy
Analytical technique used for measuring the masses of atoms or molecules in a gaseous, liquid, or solid sample

## Mutagens
Substances that tend to increase the frequency of mutation

## OECD
Organization for Economic Cooperation and Development, a Paris-based international organization created to stimulate economic progress and world trade; the convention establishing the OECD was signed in December 1960 and went into effect in September 1961; current members of the OECD are all the Western European countries, Turkey, Canada, Japan, Australia, and the United States

## Oogenesis
Formation and maturation of the egg

## Organophosphates
Organophosphorous compounds whose insecticidal and acaricidal properties are due to esters of orthophosphoric, thiophosphoric, and dithiophosphoric acids; very toxic to man and animals because they inhibit cholinesterase and other enzymes

## Pesticides
Natural, synthetic or biosynthetic chemicals used to kill weeds, insects, fungi, rodents, and other organisms that are directly or indirectly harmful to man, not including disease-bearing microorganisms

## Pheromones
Substances that are secreted by an animal and influence the behavior of other individuals of the same species

## Photochemical smog
Atmospheric pollution due to substances (oxidants) produced by a photochemical process (a process requiring sunlight); photochemical oxidants (secondary pollutants) are produced by the reaction of hydrocarbons and nitrogen oxides (primary pollutants) under the influence of ultraviolet light; two of the most harmful photochemical oxidants are ozone and peroxyacylnitrate, the simplest member of a group of compounds known as PAN (peroxyacylnitrates)

## Phytopharmacological agents
Chemicals used to protect plants from parasites or, more generally, to treat plant disease

## Phytotoxic
Poisonous to plants

## Plant hormone
An organic substance that is produced in minute amounts in one part of the plant and carried to another area where it influences a specific physiological process; also called phytohormone

## Plastics
Polymeric materials of high molecular weight that can be molded, cast, or laminated into objects, films, or filaments; generally refer to a finished product, which is treated with plasticizers, pigments, and stabilizers

## Pollution
Change in the chemical, physical, or biological characteristics of the environment (air, water, soil) that is caused directly or indirectly by man's activities and is detrimental to the health of living organisms

**Progestins**
Progesterone modifications or derivatives that act like progesterone, the hormone that mediates many processes of gestation and pregnancy

**Prostaglandins**
A group of oxygenated unsaturated cyclic fatty acids that are found in the seminal fluid and perform a variety of functions

**Proteins**
Large and very complex molecules that consist of hundreds or thousands of amino acids linked together by peptide bonds; they are essential constituents of all living cells, are synthesized from nucleic acids, and form, for example, enzymes, antibodies, and numerous hormones

**Proteolytic enzymes**
Also called proteases, a group of enzymes that initiate the hydrolysis of proteins to amino acids; some are digestive enzymes and are located on the outside of cells, while others (cathepsins) act within the cells

**Psychotropic drugs**
Drugs that influence psychic functions and modify behavior by altering the neurophysiological and biochemical activities of the central nervous system; tranquilizers, stimulants, and hallucinogens may cause drug addiction

**Raw materials**
Materials, whether crude or processed, that can be converted by processing and or manufacture into new and useful products; for example, crude oil is the raw material for the production of naphtha, which in turn is the raw material for the production of petrochemicals

**Rodenticides**
Chemicals used to kill rodents, especially mice and rats, the most modern are coumarine derivatives that act as anticoagulants at very low concentrations

**SIPRI**
Stockholm International Peace Research Institute, a foundation created in 1966 for the purpose of carrying out research into problems concerning world peace and security

**Sterilization**
The elimination of microorganisms and their spores by means of physical agents (heat, filtration, radiation): Heating is used to pre-

serve milk and other food products; filtration is used for pharmaceuticals and cosmetics; sterilization by radiation is done by employing different radiations (ultraviolet, gamma and beta rays) depending on the resistance of the microorganisms to be destroyed and on the type of material to be sterilized

**Steroids**
Large group of natural organic substances characterized by a fundamental nucleus, sterano (or cyclopentaneperhydrophenanthrene), with four rings of carbon atoms and one or more lateral chains

**Sterols**
Solid, waxy alcohols with 27–29 carbon atoms that are found in nature as esters of aliphatic acids of relatively high molecular weight; distinguished in zoosterols (animals), phytosterols (plants), mycosterols (yeasts and fungi), and sterols of marine origin (sponges)

**Sulfonamides (sulfa drugs)**
Bacteria-inhibiting compounds used in the treatment of infections caused by streptococci, pneumococci, staphylococci, gonococci and meningococci, such as meningitis, dysentery, various infections of the urinary tract, gonorrhea, and pneumonia; unlike antibiotics, they do not kill bacteria, but act by inhibiting the growth of colonies of pathogenic bacteria, leaving the host organism to destroy them subsequently

**Surfactants**
Organic compounds whose molecule contains both a hydrophobic and a hydrophilic group, enabling them to act at the interface of two immiscible phases (e.g., water and oil) by lowering their surface tension and thus to promote the dispersion of one phase into another; they find use in, for example, the textile industry (preparation of fibers, threads, and fabrics, dyeing, printing, etc.); the cosmetic industry (soaps, bubble bath products, shampoos, creams, etc.); the pharmaceutical and food industries; ironworking (as lubricants and emulsifiers); in agriculture (for the emulsification and spraying of pesticides, etc.); oil refining processes (to demulsify crude oil, for example); the production of polymers and paints

**Synergism**
In medicine, this term denotes the interaction of two drugs that results in the potentiation of the effects of one or both

**Teratogenesis**
The production of monstruous fetuses due to deviations in the em-

bryo's development for genetic or environmental reasons; teratogenicity is the ability to cause developmental malformations

**Terpenes**
Large groups of hydrocarbons found present in the vegetable realm, especially in essential oils and resins, and in the animal realm as constituents of numerous fats and oils

**Tetracyclines**
Broad-spectrum antibiotics produced by a soil actinomycete (as *Streptomyces aureofaciens*) or synthetically from chlortetracycline

**Thermal pollution**
Increase in water temperature detrimental to existing ecosystems; it is caused by the discharge into natural waters of liquids that have been used to cool down thermal stations, oil refineries, ironworks, and nuclear plants

**Tobaccoism (tobaccosis)**
Intoxication due to the excessive and prolonged use of tobacco

**Tolerance limit**
Point at, and beyond which, a chemical agent (additive, pesticide residue, etc.) becomes harmful to living organisms, established case by case on the basis of precise norms

**Tranquilizers**
Drugs used to relieve tension (e.g., chlorpromazine) or in the symptomatic treatment of common neuroses and psychosomatic disorders caused by tension and anxiety (e.g., meprobamate)

**Value added**
Difference between the total earnings of a company and the cost of raw materials, industrial aids, and services; it measures the value that the firm has "added" to the purchased materials and component parts

**Virus**
Any of a large group of submicroscopic infective agents that can only multiply in living cells and cause various important diseases in man, lower animals, and plants

**Vitamins**
Various organic substances that are essential in minute quantities to the metabolism and growth of most animals and are found in natural foodstuffs (animals and plants), but generally cannot be synthesized by animals; lack of vitamins in the organism causes characteristic syndromes, such as rickets in the case of vitamin D

**WHO**

World Health Organization, a Geneva-based specialized agency of the United Nations designed to promote international cooperation for improved health conditions

# Notes

## Chapter 2

a. *Clostridium botulinum* and *Clostridium parabotulinum* are anaerobic bacteria that elaborate a powerful toxin, botulin, which is resistant to the action of gastric juices. Only 30/1,000,000 grams (66/10,000,000 pounds) suffice to kill a mouse. The growth of botulinum spores can be inhibited by small amounts of antibiotics.

b. *Translator's note:* pH is defined as the negative logarithm of the effective hydrogenation concentration or hydrogen-ion activity in gram equivalents per liter used in expressing acidity-alkalinity on a scale from 0 to 14, with numbers greater (less) than 7 representing increasing alkalinity (acidity).

c. *Translator's note:* Coccidiostats are drugs that inhibit coccidia, protozoan parasites that cause coccidiosis, a common and serious disease of domestic animals.

## Chapter 3

a. Koch's bacillum was first stained with methyl blue, which made it stand out from the background. Subsequently Koch used vesuvine (Bismark brown R) to reveal the particles contained in the bacillum itself. This system of double staining was further developed by the neurologist F. Ziel and perfected by P. Ehrlich.

b. There are many discrepancies in the available data. This is due to the difficulty of obtaining reliable statistics and to the fact that different criteria are used in the classification of pharmaceutical products.

c. The bark and roots of *Rauwolfia* (family Apocyanaceae, tropical plants found especially in India) contain numerous alkaloids. About forty have been identified, including yohimbine and related stereoisomers, the reserpine group, and, less important, the serpentine, rescinnamine, ajmaline, and sarpagine groups. Reserpine was isolated by J. M. Müller in 1952 and prepared by stereospecific synthesis by Robert Burns Woodward in 1956.

d. Veratrum alkaloids are found in some Liliaceae, such as *Schoenocaulon officinale* (commonly known as sabadilla), *Veratrum viride, Veratrum album,* and *Veratrum grandiflorum.* Sabadilla extract, or veratrine, is a mixture of alkaloids, notably veratridine and cevadine. Veratrum rhizomes yield protoveratrines and other alkaloids of similar structure.

e. The beginning of chemotherapy actually precedes this discovery. It dates back to P. Ehrlich, who coined the term in 1891 and from 1904 to 1909 developed the first antitrypanosomal drugs. In 1916 W. Roehl synthesized Germanin, effective against sleeping sickness. In 1922 M. Khalil synthesized Fuadin, effective against leishmaniasis and venereal lymphogranuloma. In 1926 Ehrlich developed the antimalarian Plasmochin. In 1929 J. Berenblum observed that *Yperite* inhibited the growth of tumors. In 1935 the first chemotherapeutic sulfonamide was synthesized by F. Mietzsch and J. Klarer and tested by G. Domagk. In 1946 J. Lehman introduced *p*-aminosalicylic acid, the first antituberculosis drug.

f. Fleming's search for an antibacterial agent nontoxic to animal tissues had its first success in the discovery of lysozyme, which was not effective, however, against pathogens like staphylococcus and streptococcus. In 1928 he observed that a colony of staphylococcus appeared to be undergoing dissolution in the vicinity of a contaminating mold, *Penicillium notatum*. Fleming ascribed this effect to an antibacterial substance secreted by the mold. He christened it penicillin, but because of its instability he was not able to isolate it. It was not until 1939 that an Oxford team headed by H. W. Florey and E. B. Chain could obtain sufficiently pure penicillin to test it clinically. It was first tested at Oxford's Radcliffe Infirmary.

g. The first diuretics used in antiquity were plant extracts such as *spiritus juniperi* and *infusum uvae ursi*. In the sixteenth century Paracelsus recommended the use of mercury and its salts for the treatment of dropsy. Diuretic substances are known to have been used by T. Willis (potassium nitrate, 1724) and W. Withering (digitalis, 1785). Xanthine derivatives were commonly used from 1859 to 1903, and 1919 saw the introduction of organic mercurials.

h. On the basis of their effects they can be divided in three groups: psycholeptics, which induce states of sedation (e.g., tranquilizers and neuroleptic drugs); psychoanaleptics, which have a stimulating effect (e.g., amphetamines); and psychodelics, which can cause changes in normal mental activity (e.g., hallucinogens and psychotomimetics). The first synthetic psychotropic drug was a barbiturate (1903). Amphetamine (or Benzedrine) was synthesized in 1935 and was widely used as a stimulant in World War II. In 1943 A. Hoffmann accidentally discovered lysergic acid diethylamide, or LSD. In 1948 V. Erspamer found that serotonin is structurally related to several psychotropic drugs. 1954 was the year of reserpine, chlorpromazine, and meprobamate. Psychopharmacology suffered a setback in 1962 when thalidomide was found to cause phocomelia in fetuses.

i. The first antihistamine compounds are due to D. Bovet (1937). They may be divided in three groups: drugs with sympatholytic action (Antergan, developed by Malpen in 1941; Antistin; Diatrin, developed by A. Ercoli in 1948; Teoforin); drugs structurally related to parasympatholytics and spasmolytics (e.g., Benadryl, Fargan, Trimeton); and aminopyridine derivatives (Neo-Antergan, Pyribenzamine).

j. The following story may be of some interest in this regard. The Seveso tragedy has been made even worse by the deplorable fighting between abortionist and antiabortionist groups. As reported by *Il Messaggero* (3 March 1977), during a meeting organized by various Catholic organizations (Communion and Liberation, Catholic Action, and the relief chapter of the ACLI) a pharmacologist stated, among other things, that dioxin is not a substance that people should worry about too much; that what is toxic to animals is certainly not 100% toxic to humans; and that while there is proof of dioxin's toxicity for laboratory animals, we do not yet know what its effects are on man.

k. This whole story is a beautiful example of the contradictions of the modern world. The same magazine devoted an entire issue to that important South American religious movement known as "theology of liberation" [see, in particular, C. Geffré, *Concilium* 6:974 (1974)]. It contained a violent attack on European theologians, accused of supporting imperialist policies, from which I have chosen the following excerpt: "Latin American theologians . . . are even more radically opposed to the various 'progressive' theologies of the Western world . . . . They share the common defect of unwittingly playing the game of the Western capitalist societies . . . . Latin American theology condemns European theology for its alleged neutrality, which, under the pretext of maintaining the relativity of the political and the absoluteness of the religious, in effect provides ideological justification for the system established by the capitalist West." One cannot help wonder by which quirk of translation it has come about that in the very same Latin America the Vatican's position in the matter of contraceptives, widely criticized (even within the Vatican itself) as a sign of excessive conservatism, has become synonymous with antiimperialism.

## Chapter 4

a. For reasons that I am at a loss to explain, infants seem to be the habitual recipients of chlorinated compounds (whose disposal in industrial effluents, on the other hand, is subject to the most stringent regulations). At least as late as April 1978, Italian drugstores sold a preparation for the sterilization of baby bottles based on 2,4,6-trichlorophenol.

b. The definition reads as follows: "A cosmetic product means any substance or preparation intended for placing in contact with the various external parts of the human body (epidermis, hair system, nails, lips and external genital organs) or with the teeth and the mucous membranes of the oral cavity with a view exclusively or principally to cleaning them or protecting them in order to keep them in good condition, change their appearance, perfume them, or correct body odors."

## Chapter 6

a. The role of chemistry in metallurgy varies from case to case. Iron working, for example, cannot be considered a chemical process.

## Chapter 7

a. A small number of far-sighted people tried to call our attention to the dangers of such a policy. One of them is Felice Ippolito who in 1963 wrote these prophetic words in *La politica del CNEN* (Milan) concerning oil imports from the Middle East: "Such imports, affected as they are by the ups and downs of the international situation and by the instability of the Middle Eastern countries . . . are likely to trigger sudden and serious economic crises." He also wrote, "Evaluations of the competitiveness of any new energy source must take into account a coefficient expressing the probability that traditional energy sources may become scarce or wholly unavailable for technical or political reasons, as well as the serious economic consequences for a country like ours that relies so heavily on imports."

b. Representative of this attitude is the interview with Giuseppe Saragat carried by *Il Corriere della Sera* of 11 August 1963. Referring to plans for the construction of nuclear power plants, the socialist leader stated, "In Italy today there is a veritable obsession with nuclear energy . . . Building nuclear power plants for the sole purpose of generating electric energy is like building a sawmill just to produce saw dust. Sawmills are built to make boards and beams; saw dust is a by-product that, packaged and sold, helps to reduce costs. Building a sawmill to give away boards and beams for close to nothing and to sell saw dust is an absurdity."

c. Fred Hoyle, the British astronomer known in the scientific community for his theory of the expansion of the universe and to science fiction buffs for his novel *The Black Cloud,* discusses this issue in a very interesting book entitled *Energy or Extinction,* London (1977). Hoyle holds that if we wish to maintain (or raise) our current standard of living, we have no choice but to adopt the nuclear solution as quickly as possible. Opposing views are frequently expressed by knowledgeable people. See, for example, A. B. Lovins, *Energia dolce,* Milan (1979).

## Chapter 8

a. In Checkov's monologue "Is tobacco harmful?," we read, "If you put a fly in a humidor, it will die, probably from nervous shock."

b. Father Bartolomé de las Casas describes the encounter of Europeans with tobacco in his *Apologética Historia:*
In this island of Hispaniola and in the neighboring ones they had another kind of weed, similar to lettuce, that they dried in the sun and by the fire. Then, by rolling up dried leaves of trees as if they were pieces of paper, they would make small tubes, stuff them with the weed, light one end of the roll, and absorb or inhale the smoke into their chest from the other end—a thing that caused them such a torpor in their flesh and in their whole body that they would not feel hunger or fatigue. These small rolls they called *tobacos* (with the accent on the middle syllable).
    Some of our people hold that smoking these weeds is more a vice and a bad habit, or a delusion, than a real help, but they judge it so because they have no knowledge of other weeds that produce the same effects and are just as powerful . . . . I met a Spaniard, married and well respected on this island, who used *tobacos* and their smoke I mentioned above as the Indians did, and who would say that because of the great benefit he derived from them he would never give them up, even though some of his neighbors, who did not understand the benefit, condemn it as a great vice.

c. Peyote, or peyotl, is a type of cactus found in Mexico and Texas from which a drug can be obtained. It contains some alkaloids, such as mescaline and peyotine, that have strong narcotic properties. Native Indians have used peyote for hundreds of years in religious ceremonies to produce inebriation with hallucinatory visions and some local populations still eat it, after special treatment, or prepare a beverage from it.

d. As N. Kessel and H. Walton remark in *Alcoholism,* Penquin Books (1965), "Though we frown on drunkards, we are suspicious of teetotalers."

e. The piece of news reported, for example, leaves the reader with the perplexing vision of 5,000 sheep wandering alone in a Utah valley since from all accounts no shepherds appear to have been hit by a cloud that spread for 45 miles. A similar incident involving the military occurred in Switzerland in 1941. As reported by Lausanne's *Tribune-Le Matin* of 28 April 1977, "Smoke bombs poisoned 14,000 head of cattle mostly in central Switzerland. All the animals had to be slaughtered." The newspaper makes no mention of human casualties.

f. The definitions contained in the 1977 *SIPRI Yearbook* are the official definitions of the US government:

A nuclear weapon (Broken Arrow) *accident* is any unexpected event involving nuclear weapons or nuclear components which results in any of the following: (a) accidental or unauthorized launching, firing or use, by US forces or US-supported allied forces, of a nuclear–capable weapon system which could create the risk of outbreak of war; (b) nuclear detonation; (c) nonnuclear detonation/burning of a nuclear weapon; (d) radioactive contamination; (e) seizure, theft, or loss of a nuclear weapon or nuclear component, including jettisoning; and (f) public hazard, actual or implied.

A nuclear weapon (Bent Spear) *incident* is any unexpected event involving nuclear weapons or nuclear components which does not fall in the nuclear weapon accident category but which (a) results in evident damage to a nuclear weapon or nuclear component to the extent that major rework, complete replacement, or examination or recertification by the Energy Research & Development Administration (ERDA) is required; or (b) requires immediate action in the interest of safety or which may result in adverse public reaction (national or international) or premature release of information.

## Chapter 10

a. Bertrand Russell expresses this tendency in a more radical way in *Power—A New Social Analysis:* "While animals are content with existence and reproduction, men desire also to expand, and their desires in this respect are limited only by what imagination suggests as possible. Every man would like to be God, if it were possible; some few find it difficult to admit the impossibility."

b. The CNRs *Energy Report (1975)* puts it this way: "There is a great deal of talk about the possibility of solving our energy problems by a new pattern of development. Very likely the need to change our system is real and the obsolescence of the old pattern, which still holds sway in the West, may be the cause of many of the current troubles beside the energy crisis. Granted all this, the fact remains that until somebody defines the new pattern it is very difficult to analyze a mere verbal expression or estimate how much energy or research it requires."

c. I owe the following priceless story to my friend and colleague F. Cacace. The first warning against the potential dangers associated with the use of fluorocarbons came from a few scientists. One of them, pressed by Cacace, had to admit that he was using deodorants and shaving creams in spray cans containing fluorocarbons.

d. At the meeting of the ministers of Agriculture of the OECD countries held in Paris on 9–10 February 1978, the French minister stated, "We cannot simply take our surpluses (which are one of the reasons for the erosion of our farmers' income) and liquidate them at bankruptcy prices on the international markets. If we did that, the farmers of other countries would be affected and in turn would have to lower their prices. And if prices are low, it is not their fault: it is the fault of those countries that have produced too much."

e. R. Vacca writes in *Il medioevo prossimo venturo*, Milan (1971), about "leaving to professional and nonprofessional spiritual leaders the task of exhorting people to be good, even though they have been at it for thousands of years without any notable success."

f. International conferences on our most serious problems do not seem to be too successful. In its 24 May 1978 issue *Il Messaggero* observed in the matter of disarmament, "The World Disarmament Conference, in preparation for over 20 years, opened yesterday at the UN glass palace in a climate of general pessimism." On the same subject, N. Wade wrote in *Science* 200(4342): 633(1978), "The United Nations Special Session on Disarmament . . . does not have a great deal to be hopeful about." Similar skepticism and a deep bitterness were expressed by A. H. Boerma, director-general of the FAO, during the concluding session of the World Food Conference (Rome, November 1974):

Perhaps you yourselves may not have been too happy tonight upon realizing that some of the results of the conference have consisted in the creation of new committees . . . . Obviously, there remains also the criticism for the fact that, when all is said and done, the conference has accomplished nothing with regard to the immediate crisis while it is estimated that 7.5 million tons of cereals are needed in the next 6 months to take care of the needs of the 30 developing countries most seriously hit by the world's current economic situation . . . . We are thus facing an incongruity. The international community has finally faced up to the gravity of the world's food problem to the point of organizing a large conference to plan ways and means of overcoming it on a large scale in the years to come. And yet we have to fight far harder to find much simpler ways and means of facing the immediate menace that this problem poses. I became bitterly aware of this incongruity last week during a visit to Bangladesh. The sight of children in a camp in one of the worst disaster areas, piteously hanging on to life in a way to move me to tears, surrounded by the dead, provokes a sense of anger on realizing that we are still too far from the terrifying reality of the hunger and malnutrition that afflict millions of people day after day while the diplomats talk until late at night.

g. Such indications must be accurately evaluated, however, which is not easy. It has recently been reported in *Chemical and Engineering News* 56(21):27 (1978) that hamburgers contain mutagenic substances.

h. The attitude of the aveage person toward the prospect of an increasingly difficult future—not to use the word "ecocatastrophe" dear to disaster buffs—is very peculiar. Asked what he thinks about it all, he will typically answer, "Something will be done," "Things will take care of themselves," or "Science will find a way." That is, he shows a certain optimism. If we now turn from universal problems to more trivial issues and ask the same person, for example, if he thinks that the Italian government will be able to get rid of at least some of our thousand "useless agencies," he will surely answer, "Are you kidding? Not a chance!" Yet the solution of the big problems depends to a very large extent on our ability to solve the small and medium ones. And eliminating just one of our "useless agencies" implies a victory over groups whose special interests, after all, are insignificant compared to the self-interest that would be threatened by, say, any attempt to reverse the country-to-city movement or to lower the standard of living of entire populations.

# References

## Chapter 1

1. The bilingual (English and French) publications of the OECD are a precious source of data. The data reported here are taken from a periodic publication, *L'Industrie chimique 1974/75*, Paris (1976), supplemented with data from *L'Industrie chimique 1975*, Paris (1977).

2. For a thorough and accessible examination of the subject, see the article by R. Truhaut, "I pericoli dell'era chimica," *Scienza & tecnica 71* (1971).

3. Council on Environmental Quality, *Environmental Quality: Report VI*, Washington, DC (1975).

4. P. Greenwald, N. J., Vianna, P. C. Nasca, and J. N. P. Davies, *Environmental Influences on the Development of Cancer*.

5. "Study urged on multiple factors causing cancer," *Chemical and Engineering News* 54(14):4 (1976).

## Chapter 2

1. The quotation is from Glenn T. Seaborg's opening speech at the 1976 meeting of the American Chemical Society. See also S. Marei "Risorse, popolazione, produzione agricola," *Scienza & tecnica* 76 (1976); H. Linneman, "Prospettive in agricoltura: Moira, uno studio mediante modello," *Scienza & tecnica* 77 (1977).

2. A. Quispel, "Incremento dell'uso di azoto atmosferico in agricoltura," *Scienza & tecnica* 77 (1977).

3. *Compendio statistico Aschimici 1976*, Milan (1976).

4. Report of the President's Advisory Committee, prepared by the Commission on World Food Sources, Washington (1967).

5. E. V. Anderson, "Steady growth pegged for pesticide use," *Chemical and Engineering News* 53(20):15 (1975).

6. R. Carson, *Silent Spring*, Boston (1962).

7. I. Illich, *Deschooling Society*, London (1971).

8. I. Illich, *Limits to Medicine—Medical Nemesis: The Expropriation of Health*, London (1976).

9. P. de Pietri-Tonelli and P. Piccardi, "I nuovi orientamenti nella lotta contro gli insetti," *Scienza & tecnica* 76 (1976).

10. For additional information on the difficulties of storing agricultural goods see D. W. Hall, *Manutention et emmagasinage des grains alimentaires dans les régions tropicales et subtropicales,* Collection FAO number 90, Paris (1971); Abdel-Hamid and F. Abdel-Aziz, *Emmagasinage des grains alimentaires,* Paris (1976).

11. National Academy of Sciences, *Pest Control and Public Health,* V, Washington, DC (1976).

12. A thorough examination of the Seveso incident in all its aspects can be found in *Sapere* 79:796 (1976). See also L. Canonica, "Seveso: considerazioni e commenti," *La Chimica e L'Industria* 59(2):86 (1977).

13. *Food and Nutrition* 2(3):4 (1976).

14. "FDA to classify saccharin as a drug," *Chemical and Engineering News* 55(16):8 (1977).

15. "Saccharin: A chemical in search of an identity," *Science* 196(4295):1179 (1977).

16. For this analysis see the letter by D. S. Matteson and the literature cited in it, *Chemical and Engineering News* 55(18):4 (1977).

17. W. Lijinski, "U.S. health will be jeopardized if Delaney clause is abandoned," *Chemical and Engineering News* 55(26):27 (1977).

18. E. M. Boyd, *Toxicity of Pure Foods,* Cleveland (1973).

**Chapter 3**

1. I. Illich, *Limits to medicine—Medical Nemesis: The Expropriation of Health,* London (1976).

2. American Chemical Society, *Chemistry and Economy,* Washington, DC (1973).

3. D. Derrick, "Farmaci e malattie iatrogene," *Scienza & tecnica, 74* (1974).

4. V. Andreoli, F. Maffei, and G. Tamburino, *Il ciclo della droga,* Milan (1978).

5. C. Djerassi, "Prognosis for the development of new chemical birth-control agents," *Science,* 160:468, 945 (1970). See also Djerassi, *Science* 151:1055 (1966) and 156:472 (1969).

6. V. Beral, "Mortality among oral-contraceptive users," *The Lancet* 8041:727 (1977).

7. V. Cosmao, "Regolazione demografica e sviluppo," *Concilium* 10:91 (1975).

8. N. Aspesi, "Per i baroni della pillola le donne non esistono," *La Repubblica* (8 March 1977).

9. A. Burger, "Behind the decline in new drugs," *Chemical and Engineering News* 53(38):37 (1975).

**Chapter 4**

1. American Chemical Society, *Chemistry and Economy,* Washington, DC (1973).

2. *Pest. Chem. News* 1(23):3 (1973). For a thorough exploration of the chemical aspects of cosmetology see R. Selleri, C. Botré, and G. Orzalesi, *Chimica e tecnologia dei prodotti cosmetici,* Rome (1977). Concerning the dangers of cos-

metology see C. N. Roberts, "La sicurezza nell'uso dei cosmetici," *Scienza & tecnica* 76 (1976).

## Chapter 5

1. An accurate summary of the chemistry and technology of plastic materials is given in E. Cernia, "Alti polimeri: plastomeri, elastomeri, fibre," *Enciclopedia della Chimica,* IV, Milan (1977).

2. Useful information about the most important chemical products of the plastics industry (and other industries) can be found in the booklet *Key Chemicals,* published in 1978 by *Chemical and Engineering News.*

3. For more on this subject and for the treatment of solid urban refuse, see W. Ganapini, *La risorsa rifiuti,* Milan (1978).

4. For a thorough special report on fire retardants, see H. J. Sanders, "Flame retardants," *Chemical and Engineering News* 56(17):22 (1978).

5. Uses and drawbacks of fluorocarbons in spray cans are given in W. R. Muir and C. L. Pegler, "Fluorocarburi e ambiente," *Scienza & tecnica* 76 (1976).

6. M. J. Molina and F. S. Rowland, "Stratospheric sink for chlorofluoromethanes: Chlorine atom-catalysed destruction of ozone," *Nature* 249(5460):810 (1974).

7. Works by J. Ausloos and R. E. Rebbert of the National Bureau of Standards cited in the 28 February 1977 issue of *Chemical and Engineering News.*

8. Council on Environmental Quality, *Environmental Quality: Report VIII,* Washington, DC (1977).

## Chapter 6

1. G. Costa and G. Mestroni, "I simulatori enzimatici," *Scienza & tecnica* 78 (1978); A. Quispel, "Incremento dell'uso dell 'azoto atmosferico in agricoltura," *Scienza & tecnica* 77 (1977).

2. M. Conserva and F. Gatto, "Leghe di alluminio: impieghi e tecnologie," *Scienza & tecnica* 78 (1978).

3. The depletion of mineral resources is treated in D. Gabor, U. Colombo, A. King, and R. Galli, *Oltre l'età dello spreco,* Milan (1976); MIT Club of Rome, *I limiti dello sviluppo,* Milan (1974).

4. F. Ahmed, "Les métaux et l'avenir du monde," *La Recherche* 73:1046 (1976).

5. E. Davin, "Materie prime dal fondo oceanico," *Scienza & tecnica* 77 (1977).

6. Swedish Ministries of Exterior and Agriculture, *The Impact on the Environment of Sulfur in Air and Precipitation,* a study presented at the UN–sponsored 1972 Stockholm Conference on the Environment.

7. J. M. Wood, "Les métaux toxiques dans l'environment," *La Recherche* 70:711 (1976).

8. J. Aulka and T. H. Risby, *Analytical Chemistry* 48:640 (1976).

9. W. P. Ridley, L. J. Dizikes, and J. M. Wood, "Biomethylation of toxic elements in the environment," *Science* 197:329 (1977).

10. H. Péquignot, "Le malattie del mondo moderno," *Scienza & tecnica* 74 (1974).

11. P. Pons, "Les malades honteux de Minamata," *Le Monde* (23 March 1978).

12. W. B. Deichman and H. W. Gerade, *Toxicology of Drugs and Chemicals,* New York (1969).

13. *Report on Third Technical Meeting on Occurrence and Significance of Chemicals in the Environment,* Paris (1972), from a meeting held in Berlin in 1972.

14. L. Favretto, G. Pertoldi and G. Favretto, *Piombo nell'ambiente,* Udine (1975). This work was sponsored by the University of Trieste.

15. E. J. Underwood, *Trace Elements in Human and Animal Nutrition,* New York (1971).

**Chapter 7**

1. D. Gabor, U. Colombo, A. King, and R. Galli, *Oltre l'età dello spreco,* Milan (1976). This work contains accurate information on the production of energy and its importance in materials processing.

2. From a report presented by E. Cook at the meeting of the AIMME (American Institute of Mining and Metallurgical Engineers) held in Dallas on 26 February 1974.

3. E. T. Hayes, *Energy Implications of Materials Processing,* in P. H. Abelson and A. L. Hammond, eds., *Materials: Renewable and Nonrenewable Resources,* Washington, DC (1976), a special issue of *Science* published by the American Association for the Advancement of Science.

4. The ratio between yield and energy expenditure in various agricultural systems is given in G. Leach, *Energy and Food Production,* London (1975).

5. M. Rawitscher and J. Mayer, "Nutritional outputs and energy inputs in seafoods," *Science* 198:261 (1977).

6. On the question of energy and food, see the book by Gabor et al. cited in note 1, and also see 'FAO, *La situation mondiale de l'alimentation et de l'agriculture,* Paris (1976).

7. The study, by Arthur D. Little Inc., is cited in R. Seymour and J. M. Sosa, "Plastics from plastics," *Chemtech* 7(8):507 (1977).

8. N. Keyfitz, "Risorse mondiali e ceto medio," *Le Scienze* 17(99):13 (1976).

9. P. H. Abelson, "Foreword" in *Energy: Use, Conservation and Supply,* Washington, DC (1974).

10. These two studies are reported in the OECD publication *La production d'énergie et l'environment,* Paris (1977).

11. Arthur D. Little Inc., *Combating Pollution Created by Oil Spills* (1969). The data are quoted in OECD, *La production d'énergie et l'environment,* Paris (1977).

12. For the problems associated with strip mining, see M. Waldrop, "Strict law challenges strip mine operators," *Chemical and Engineering News* 55(34):18 (1977).

13. American Chemical Society, *Trace Elements in Fuels,* Washington (1975).

14. E. Piperno, "Trace element emissions: Aspects of environmental toxicology," in American Chemical Society, *Trace Elements in Fuels,* Washington, DC (1975).

15. C. E. Chrisp, G. L. Fisher, and J. E. Lammert, "Mutagenicity of filtrates from respirable coal fly ash," *Science* 199(4324):73 (1978).

16. L. J. Carter, "Coal: "Invoking 'the rule of reason' in an energy-environment conflict," *Science* 198(4314):276 (1977).

17. E. Gruner, *Inst. Civil Eng. Proc.* 24:47 (1963).

18. G. A. Kiersch, *Geotimes* 34:12 (1965).

19. V. Gilincki, *Chemical and Engineering News* 55:18 (1977).

20. M. Benedict, "The safety of nuclear power," *Chemical and Engineering News,* 55(28):5 (1977).

21. In P. H. Abelson, *Energy: Use, Conservation and Supply,* Washington, DC (1974).

22. "Le fonti alternative di energia," *Scienza & tecnica* 76 (1976).

**Chapter 8**

1. The quotation is taken from Cournand A., "The code of the scientist and its relationship to ethics," *Science* 198(4318):699 (1977).

2. Among the articles or essays that foresaw the great changes that are now obvious to everybody are J. Platt, "What we must do," *Science* 166:(1115) (1969); R. Vacca, *Il medioevo prossimo venturo,* Milan (1971).

3. P. E. James, *Latin America,* New York (1959).

4. M. Silvestri, *La decadenza dell'Europa occidentale,* Turin (1977).

5. *Panorama* (27 June 1978).

6. *Euroforum* 38 (25 October 1977).

7. G. Bonnot, *La vie c'est autre chose,* Paris (1976).

8. G. Menahem, *La science et le militaire,* Paris (1976).

9. Nguyên Dang Tâm, "Le Vietnam, champ d'expérience pour la guerre chimique," *La Recherche* 1:442 (1970).

10. C. Brisset, *Le Monde* (4 August 1977). The news report was taken from the *New York Times* (2 August 1977).

**Chapter 9**

1. C. Starr, R. Rudman, and C. Whipple, "Philosophical basis for risk analysis," *Annual Review of Energy* 1 (1976).

2. A. Plant, "A lot of results, but not many answers," *Chemical and Engineering News,* 54(40):2 (1976).

3. C. Comar, "Environmental assessment: A Pragmatic View," *Science* 198(4317):567 (1977).

4. A. Sampaolo, *Rassegna Chimica* 29:219 (1977).

5. C. Murray, "Chemical firms wary over toxic substances law," *Chemical and Engineering News* 55(1):15 (1977).

6. H. J. Sanders, "Chemical lab safety and the impact of OSHA," *Chemical and Engineering News* 54(22):15 (1976).

**Chapter 10**

1. N. Keyfitz, "World resources and the world middle class," *Scientific American,* 235(1):28 (1976).

2. M. J. Williams, "Satisfaire les besoins essentiels des populations les plus pauvres," *L'Observateur de l'OCDE* 89:17 (1977).

3. J. H. Knowles, "Responsibility for health," *Science* 198(4322):1103 (1977) [adapted from his article in *Daedalus* 106:57 (1977); reprinted in *Doing Better and Feeling Worse: Health in the United States,* New York (1977), 57].

4. R. J. Smith, "Uncle Sam wants you to be healthy and inexpensive," *Science* 200(4338):186 (1978).

5. D. J. Rose, in P. H. Abelson, *Energy: Use, Conservation and Supply,* Washington, DC (1974).

# Bibliography

**Chapter 1**
General references on the relation between chemistry and society and on the development of the chemical industry are

M. M. Jones, J. T. Netterville, D. O. Johnston, and J. L. Wood, *Chemistry, Man and Society* (New York: Saunders, 1980)

OECD, *The Chemical Industry* (Paris: OECD, 1979)

B. G. Reuben and M. L. Burstall, *The Chemical Economy* (London: Longmans, 1978)

Royal Society of Chemistry, *Chemistry and the Needs of Society* (London: The Royal Society of Chemistry, 1974)

W. A. Wittcoff and B. G. Reuben, *Industrial Organic Chemicals in Perspective*, parts 1–2 (New York: Wiley, 1980)

J. H. Woodburn, *Taking Things Apart and Putting Things Together* (Washington, DC: American Chemical Society, 1976)

More detailed information on chemistry and the environment, toxicity of chemicals (for example, vinyl chloride and asbestos), and chemistry and human health may be found in

R. A. Bailey, H. M. Clarke, J. P. Ferris, S. Krause, and R. L. Strong, *Chemistry of the Environment* (New York: Academic Press, 1978)

K. Berg, ed., *Genetic Damage in Man Caused by Environmental Agents* (New York: Academic Press, 1979)

M. E. Green and A. Turk, *Safety in Working with Chemicals* (New York: Macmillan, 1978)

M. Lippmann and R. B. Schlesinger, *Chemical Contamination in the Human Environment* (Oxford: Oxford University Press, 1979)

G. S. Rajhans and B. M. Bragg, *The Engineering Aspects of Asbestos Dust Control* (New York: Wiley, 1978)

I. J. Selikoff and D. H. K. Lee, *Asbestos and Disease* (New York: Academic Press, 1978)

M. Sitting, *Vinyl Chloride and PVC Manufacture—Process and Environmental Aspects* (Noyes Data Corporation, 1978)

Up-to-date statistics on causes of death are in

M. Anderson, *International Mortality Statistics* (New York: Macmillan, annual)

## Chapter 2

For general information on economy, technology, and the needs of agricultural industry, consult

ACS Committee on Chemistry and Public Affairs, *Chemistry and the Food System* (Washington, DC: American Chemical Society, 1980)

I. Arnon, *Modernization of Agriculture in Developing Countries* (New York: Pergamon, 1980)

M. Chou and D. Harmor, Jr., eds., *Critical Food Issues of the Eighties* (New York: Pergamon, 1979)

G. W. Cooke, ed., *Agriculture Research 1931–1981: A History of the Agricultural Research Council and a Review of Developments in Agricultural Science during the Last Fifty Years* (London: Agricultural Research Council, 1981)

G. L. Cramer and C. W. Jensen, *Agricultural Economics and Ecological Comparison with Conventional Methods* (New York: Wiley, 1979)

D. C. Dahl and J. Hammond, *Market and Price Analysis: The Agricultural Industries* (New York: McGraw-Hill, 1977)

K. A. Dahlberg, *Beyond the Green Revolution* (New York: Plenum, 1979)

D. Ensminger, ed., *Food Enough or Starvation for the Millions* (New York: McGraw-Hill, 1977)

FAO, *Fourth World Food Survey* (United Nations, 1977)

N. M. Idaikkadar, *Agricultural Statistics* (New York: Pergamon, 1979)

R. C. Oelhaf, *Organic Agriculture: Economic and Ecological Comparison with Conventional Methods* (New York: Wiley, 1979)

*The Political Economy of Agrarian Change: An Essay on the Green Revolution* (Cambridge, MA: Harvard University Press, 1974)

*Technology, Employment and Basic Needs in Food Processing in Developing Countries* (New York: Pergamon, 1980)

Up-to-date accounts of chemical involvement in the agricultural uses of fertilizers are

C. W. Fowler, *Urea and Urea Phosphate Fertilizers* (Noyes Data Corporation, 1976)

P. A. Hendrie, *Granulated Fertilizers* (Noyes Data Corporation, 1976)

National Research Council, *World Food and Nutrition Study* (Washington, DC: National Academy of Sciences, 1977)

N. W. Pirie, *Food Resources* (New York: Penguin Books, 1976)

Royal Academy of Chemistry, *Chemistry and Agriculture* (London: The Royal Academy of Chemistry, 1979)

M. Sittig, *Fertilizer Industry* (Noyes Data Corporation, 1979)

General information on pesticides—their use, nature, modes of action, toxicity, persistence, economics, benefits, and risks—may be found in

Advisory Committee on Pesticides, *Further Review of the Safety for Use in the U.K. of the Herbicide 2,4,5-T* (Amersham: Robendene Ltd. for Her Majesty's Stationery Office, 1980)

American Chemical Society, *Controlled Release Pesticides* (Washington, DC: American Chemical Society, 1977)

F. M. Ashton and A. S. Crafts, *Mode of Action of Herbicides*, 2nd ed. (New York: Wiley, 1981)

M. Beroza, ed., *Pest Management and Sex Attractants* (Washington, DC: American Chemical Society, 1976)

E. H. Blair, "The safety of 2,4,5-T," *Science* 206:135 (1979)

G. T. Brooks, *Chlorinated Insecticides*, vols. 1-2 (Boca Raton: CRC Press, 1974)

A. W. A. Brown, *Ecology of Pesticides* (New York: Wiley, 1978)

J. E. Casida, ed., *Eyrethrum: The Natural Insecticide* (New York: Academic Press, 1973)

Commission du Codex Alimentarius, *Guide Concernant les Limites Maximales Codex pour Residues de Pesticides* (FAO/OMS, CAC/PRI, 1978)

Commission of the European Communities, *Criteria (Dose/Effect Relationships) for Humans on Organochloride Compounds* (New York: Pergamon Press, 1979)

R. Cremlyn, *Pesticides: Preparation and Mode of Action* (New York: Wiley, 1978)

T. R. Dunlap, *DDT: Scientists, Citizens and Public Policy* (Princeton: Princeton University Press, 1981)

C. E. Edwards, *Persistent Pesticides in the Environment* (Boca Raton: CRC Press, 1974)

M. Eto, *Organophosphorus Pesticides: Organic and Biological Chemistry* (Boca Raton: CRC Press, 1974)

H. Freshse and H. Geissbuhler, eds., *Pesticide Residues* (New York: Pergamon Press 1979)

H. Geissbuhler, ed., *World Food Production–Environment–Pesticides* (New York: Pergamon Press, 1979)

L. I. Gilbert, *The Juvenile Hormones* (New York: Plenum Press, 1976)

M. B. Green, G. S. Hartley, and T. F. West, *Chemicals for Crop Protection and Pest Control* (New York: Pergamon Press, 1977)

G. S. Hartley and I. J. Graham-Bryce, eds., *Physical Principles of Pesticide Behavior*, vol. 2 (New York: Academic Press, 1980)

F. L. McEwen and G. R. Stephanson, *The Use of Significance of Pesticides in the Environment* (New York: Wiley-Interscience, 1978)

N. R. McFarlance, *Herbicides and Fungicides: Factors Affecting Their Activity* (London: The Royal Academy of Chemistry, 1977)

National Research Council, *Polichlorinated Biphenyls* (Washington, DC: National Academy of Sciences, 1979)

National Research Council of Canada, *Phenoxy Herbicides–Their Effects on Environmental Quality* (NRCC no. 16075, 1978)

Office of Technology Assessment, *Environmental Contaminants in Food* (Washington, DC: US Government Printing Office, 1979)

J. Perfect, "The environmental impacts of DDT in a tropical agro-ecosystem," *Ambio* 9:16 (1980)

F. H. Perring and K. Mellanby, eds., *Ecological Effects of Pesticides* (New York: Academic Press, 1977)

D. Pimental, ed., *Handbook of Pest Management in Agriculture*, vol. 2 (Boca Raton: CRC Press, 1981)

J. R. Plimmer, ed., *Pesticides Chemistry in the 20th Century* (Washington, DC: American Chemical Society)

S. S. Que Hee and R. G. Sutherland, *The Phenoxyalkanoic Herbicides*, vol. 1: *Chemistry, Analysis and Environmental Pollution* (Boca Raton: CRC Press, 1981)

F. J. Ritter, ed., *Chemical Ecology: Odour Communication in Animals* (Amsterdam: Elsevier/North Holland Biomedical Press, 1979)

M. Sittig, *Pesticide Manufacturing and Toxic Materials Control Encyclopedia* (Noyes Data Corporation, 1980)

K. Slama, M. Romanuk, and F. Sorm, *Insect Hormones and Bioanalogues* (New York: Springer-Verlag, 1974)

E. H. Smith and D. Pimentel eds., *Pest Control Strategies* (New York: Academic Press 1978)

L. A. Summers, *Bipyridinium Herbicides* (New York: Academic Press, 1980)

G. Vettorazzi, *International Regulatory Aspects for Pesticide Chemicals*, vol. 1: *Toxicity Profiles* (Boca Raton: CRC Press, 1979)

D. L. Watson and A. W. A. Brown, eds., *Pesticide Management and Insecticide Resistance* (New York: Academic Press, 1977)

WHO Working Group, *Toxicological Appraisal of Halogenated Compounds Following Groundwater Pollution* (Copenhagen: WHO 1980)

C. R. Worthing, ed., *The Pesticide Manual* (London: British Crop Protection Council, 1979)

Treatments of the Seveso accident are found in

E. H. Blair, ed., *Chlorodioxins—Origins and Fate* (Adv. Chem. Ser. 120, 1973)

Health and Safety Executive, *English Translation of the Final Report of the Italian Parliamentary Commission of Enquiry on the Escape of Toxic Substances on 10 July 1976 at the ICMESA Establishment: Seveso* (London: 1980)

T. Whiteside, *The Pendulum and the Toxic Cloud—The Course of Dioxin Contamination* (New Haven: Yale University Press, 1979)

Further entries on the scientific principles, practical techniques, and risks of food conservation by nonchemical and chemical methods are

F. Coulston, ed., *Regulatory Aspects of Carcinogenesis and Food Additives* (New York: Academic Press, 1980)

T. R. Crompton, *Additive Migration from Plastics into Food* (New York: Pergamon Press, 1979)

Food Safety Council, *Proposed System for Food Safety Assessment* (New York: Pergamon Press, 1978)

T. E. Furia, ed., *Handbook of Food Additives*, 2nd ed. (Boca Raton: CRC Press, vol. 1 1973, vol. 2 1980)

T. E. Furia, *Regulatory Status of Direct Food Additives* (Boca Raton: CRC Press, 1980)

T. E. Furia and N. Bellanca, eds., *CRC Fenaroli's Handbook of Flavor Ingredients* vols. 1–2 (Cleveland: CRC Press, 1975)

M. H. Gulcho, *Freeze Drying Processes for the Food Industry* (Noyes Data Corporation, 1977)

R. K. Gurthie, *Food Sanitation*, 2nd ed. (Westport: AVI, 1980)

D. Hahn-Astre, *Repertoire des Colorants Usuels: 1 Colorants à Usage Alimentaire* (Paris: Editions SCM)

International Agency for Research on Cancer, *Some Non-Nutritive Sweetening Agents* (Lyons: 1980)

J. C. Johnson, *Emulsifiers and Emulsifying Techniques* (Noyes Data Corporation, 1976)

A. Y. Leung, *Encyclopedia of Common Natural Ingredients Used in Foods, Drugs, and Cosmetics* (New York: Wiley-Interscience, 1980)

P. H. Li and A Sakai, eds., *Plant Cold Hardiness and Freezing Stress: Mechanisms and Crop Implications* (New York: Academic Press, 1978)

D. M. Marmion, *Handbook of US Colorants for Foods, Drugs, and Cosmetics* (New York: Wiley, 1979)

M. J. Nash, *Crop Conservation and Storage* (New York: Pergamon Press, 1979)

S. J. Palling, *Developments in Food Packaging* (Barking: Applied Science, 1980)

N. D. Pintauro, *Food Flavoring Processes* (Noyes Data Corporation, 1976)

I.A. Taub, R. A. Kaprielian, and J. W. Halliday, *Food Preservation by Irradiation* (Vienna: IAEA, 1978)

R. J. Taylor, *Food Additives* (New York, Wiley-Interscience, 1980)

The following books treat the utilization of residues and by-products for animal nutrition:

A. Champagnat and T. Adrian, *Petrole et Proteines* (Paris: DOIN, 1974)

P. Devis, ed., *Single Cell Protein* (New York: Academic Press, 1974)

J. T. Huber, ed., *Upgrading Residues and Byproducts for Animals* (Boca Raton; CRC Press, 1981)

P. J. Rockwell, *Single Cell Proteins from Cellulose and Hydrocarbons* (Noyes Data Corporation, 1976)

## Chapter 3
Works dealing with the general concerns of this chapter are

*Plagues and Society*

M. Anderson, *International Mortality Statistics* (New York: Macmillan, annual)

W. H. McNeill, *Plagues and Peoples* (New York: Doubleday, 1976)

*Pharmaceutical Industry*

C. R. Buncher and J. T. Tsay, *Statistics in the Pharmaceutical Industry* (New York: Marcel Dekker, 1981)

B. G. Reuben and M. L. Burstall, *The Chemical Economy* (London: Longmans, 1979)

*Pharmacology*

G. S. Avery, ed., *Drug Treatment Principles and Practice of Clinical Pharmacology and Therapeutics* (New York: Adis Press, 1980)

C. Benzold, *The Future of Pharmaceuticals: The Changing Environment for New Drugs* (New York: Wiley, 1981)

T. Z. Csaky, *Cutting's Handbook of Pharmacology* 6th ed. (New York: Appleton-Century-Crofts, 1979)

T. Z. Csaky, *Introduction to General Pharmacology* (New York: Appleton-Century-Crofts, 1979)

A. Y. Leung, *Encyclopedia of Common Natural Ingredients Used in Foods, Drugs, and Cosmetics* (New York: Wiley-Interscience, 1980)

J. R. Vane and S. H. Ferreira, eds., *Handbook of Experimental Pharmacology* (New York: Springer, 1979)

*Pharmaceuticals*

M. J. Antonaccio, ed., *Cardiovascular Pharmacology* (New York: Raven Press, 1980)

G. S. Banks and G. T. Rhodes, *Modern Pharmaceutics* (New York: Marcel Dekker, 1979)

S. K. Carter, M. T. Bakowski, and K. Hellerman, *Chemotherapy of Cancer*, 2nd ed. (New York: Wiley, 1981)

E. Chain, "Penicillin—the crucial experiment," *Chemtech* 10:474 (1980)

E. F. Gale, E. Cundliffe, P. E. Reynolds, M. H. Richmond, and M. J. Waring, *The Molecular Basis of Antibiotic Action* (New York: Wiley, 1981)

R. F. Gould, ed., *Diuretic Agents* (Washington, DC: American Chemical Society, 1978)

R. J. Hegyeli, ed., *Prostaglandins and Cardiovascular Disease* (New York: Raven Press, 1980)

T. Kamiya and J. Elks, eds., *Recent Advances in the Chemistry of $\beta$-Lactam Antibiotics* (London: The Chemical Society, 1977)

T. H. Maugh II, "A new wave of antibiotics," *Science* 214:1225 (1981)

B. Samuelsson, P. W. Ramwell, and R. Paoletti, eds., *Advances in Prostaglandine and Thromboxane Research* (New York: Raven Press, 1980)

A. Scriabine, ed., *Pharmacology and Antihypertensive Drugs* (New York: Raven Press, 1980)

D. A. Stringfellow, *Interferon and Interferon Inducers* (New York: Marcel Dekker, 1980)

D. L. Temple, Jr., ed., *Drugs Affecting the Respiratory System* (Washington, DC: American Chemical Society, ACS Symposium No. 118)

*Iatrogenesis*

M. Davis, J. M. Tredeger, and R. Williams, eds., *Drug Reactions and the Liver* (London: Pitman Books, 1981)

*Drug Dependence*

P. L. Broadhurst, *Drugs and the Inheritance of Behavior* (New York: Plenum, 1978)

J. Caldwell, *Amphetamines and Related Stimulants: Chemical, Biological, Clinical, and Sociological Aspects* (Boca Raton: CRC Press, 1980)

S. Cohen, *The Drug Dilemma*, 2nd ed. (New York: McGraw-Hill, 1976)

L. L. Iversen, S. D. Iversen, and S. H. Snyder, eds., *Handbook of Psychopharmacology*, sect. 1, vol. 12 (New York: Plenum, 1978)

M. J. Kreek, L. C. Gutjhar, J. W. Garfield, D. V. Bowen, and F. H. Field, "Drug interactions with methadone," *Ann. N.Y. Acad. Sci.* 261 (1976)

M. A. Lipton, A. Di Mascio, and K. F. Killam, eds. *Psychopharmacology: A Generation of Progress* (New York: Raven Press, 1979)

J. S. Madden, R. Walker, and W. H. Kenyon, *Aspects of Alcoholism and Drug Dependence* (London: Pitman Books, 1980)

S. Joseph Mule, *Cocaine: Chemical, Biological, Clinical, Social, and Treatment Aspects* (Boca Raton: CRC Press, 1976)

R. G. Newman, in collaboration with M. S. Cates, *Methadone Treatment in Narcotic Addiction: Program Management, Findings, and Prospects for the Future* (New York: Academic Press, 1977)

A. Schecter, ed., *Treatment Aspects of Drug Dependence* (Boca Raton: CRC Press, 1978)

A. Wikler, *Opioid Dependence—Mechanism and Treatment* (New York: Plenum, 1980)

*Contraceptives*

M. C. Diamond and C. C. Korenbrot, eds., *Hormonal Contraceptives, Estrogens and Human Welfare* (New York: Academic Press, 1978)

C. Djerassi, *The Politics of Contraception* (New York: W. W. Norton, 1979)

C. R. Kay, ed., *Oral Contraceptives and Health* (London: Pitman Books, 1974)

V. Petrow, "The chemistry of contraceptives", *Chemtech* 7:563 (1977)

**Chapter 4**

The following titles focus on the fundamentals of the modern cosmetic industry:

M. S. Balsam and E. Sagarin, *Cosmetics: Science and Technology* vols. 1–3 (New York: Wiley)

M. Billot and F. V. Wells, *Perfumery Technology: Art, Science, Industry* (New York: Wiley, 1976)

J. S. Jellinek, *Formulation and Function of Cosmetics* (New York: Wiley, 1971)

A. Y. Leung, *Encyclopedia of Common Natural Ingredients Used in Food, Drugs, and Cosmetics* (New York: Wiley-Interscience, 1980)

D. M. Marmion, *Handbook of U.S. Colorants for Food, Drugs and Cosmetics* (New York: Wiley, 1970)

## Chapter 5

Among the many works on polymers and new materials derived from oil processing, the following supply the reader with comprehensive overviews:

*Encyclopedia of Polymer Science and Technology* (New York: Wiley-Interscience, 1969)

H. Mark, "Polymer chemistry in Europe and America—how it all began," *J. Chem Educ.* 58:527 (1981)

C. S. Marvel, "The development of polymer chemistry in America—the early days," *J. Chem. Educ.* 58:535 (1981)

Y. L. Meltzer, *Water-Soluble Polymers—Recent Developments* (Noyes Data Corporation, 1979)

R. V. Milby, *Plastic Technology* (New York: McGraw-Hill, 1973)

M. W. Ranney, *Silicones*, vols. 1–2 (Noyes Data Corporation, 1977)

B. G. Reuben and M. L. Burstall, *The Chemical Economy* (London:Longmans, 1978)

R. B. Seymour and C. E. Carraher, Jr., *Polymer Chemistry: An Introduction* (New York: Marcel Dekker, 1981)

A. V. Tobolsky and H. F. Mark, eds., *Polymer Science and Materials* (Robert Krieger Publ. Co., 1980)

A. L. Waddams, *Chemicals and Petroleum—an Introductory Survey* (London: John Murray, 1978)

K. Weissermel and H. J. Arpe, *Industrial Organic Chemistry* (Weinheim and New York: Verlag Chemie, 1978)

A. Yehaskel, *Fire and Flame Retardant Polymers* (Noyes Data Corporation, 1979)

On specific topics, consult the following:

*Fibers*

M. E. Carter, *Essential Fiber Chemistry* (New York: Marcel Dekker, 1971)

S. G. Cooper, *The Textile Industry, Environmental Control and Energy Conservation* (Noyes Data Corporation, 1978)

H. F. Mark, S. M. Atlas, and E. Cernia, eds., *Man Made Fibers* (New York: Wiley-Interscience, 1967)

J. E. McIntyre, *The Chemistry of Fibers* (London: E. Arnold, 1972)

J. W. Palmer, *Textile Processing and Finishing Aids—Recent Advances* (Noyes Data Corporation, 1977)

J. S. Robinson, ed., *Fiber-Forming Polymers* (Noyes Data Corporation, 1980)

*Adhesives*

D. L. Bateman, *Hot Melt Adhesives*, 3rd ed. (Noyes Data Corporation, 1978)

H. R. Dunning, *Pressure Sensitive Adhesives—Formulations and Technology*, 2nd ed. (Noyes Data Corporation, 1977)

E. W. Flick, *Adhesive and Sealant Compounds and Their Formulations* (Noyes Data Corporation, 1978)

M. W. Ranney, *Epoxy Resins and Products* (Noyes Data Corporation, 1977)

S. Torrey, ed., *Adhesive Technology* (Noyes Data Corporation, 1980)

*Paints*

J. I. Duffy, *Printing Inks—Developments since 1975* (Noyes Data Corporation, 1979)

E. W. Flick, *Solvent-Based Paint Formulations* (Noyes Data Corporation, 1977)

E. W. Flick, *Interior Water-Based Trade Paint Formulations* (Noyes Data Corporation, 1980)

Noyes Data Corporation, *Water-Based Industrial Finishes—Recent Developments* (Noyes Data Corporation, 1980)

G. B. Rothenberg, *Paint Additives—Recent Developments* (Noyes Data Corporation, 1978)

A. Werner, "Synthetic materials in art conservation," *J. Chem. Educ.* 58:321 (1981)

*Soaps and Detergents*

W. C. Cutler and R. C. Davis, *Detergent Theory and Test Methods* (New York: Marcel Dekker, 1972)

A. Davidson and B. M. Milwidsky, *Synthetic Detergents* (London: Hill, 1972)

M. I. Rosen and H. A. Goldsmith, *Systematic Analysis of Surface-Active Agents* (New York: Wiley, 1972)

A. Schwartz, J. Perry, and J. Berch, *Surface Active Agents and Detergents* (New York: Interscience, 1966)

M. Sittig, *Detergent Manufacture including Zeolite Builders and Other New Methods* (Noyes Data Corporation, 1979)

R. D. Swisher, *Surfactant Biodegradation* (New York: Marcel Dekker, 1970)

*Fluorocarbons*

R. E. Banks, ed., *Organofluorine Chemicals and Their Industrial Applications* (New York: Wiley-Halsted, 1979)

A. K. Biswas, ed., *The Ozone Layer* (New York: Pergamon Press, 1979)

S. A. W. Gerstl, A. Zardecki, and H. L. Wiser, "Biologically damaging radiation amplified by ozone depletion," *Nature* 294:352 (1981)

National Academy of Sciences, *Protection against Depletion of Stratospheric Ozone by Chlorofluorocarbons* (Washington, DC: National Academy of Sciences, 1979)

National Academy of Sciences, *Stratospheric Ozone Depletion by Halocarbons: Chemistry and Transport* (Washington, DC: National Academy of Sciences, 1979)

*Toxicity*

P. N. Cheremisinoff and A. C. Morresi, *Benzene: Basic and Hazardous Properties* (New York: Marcel Dekker, 1979)

M. E. Green and A. Turk, *Safety in Working with Chemicals* (New York: Macmillan, 1978)

M. Sittig, *Vinyl Chloride and PVC Manufacture* (Noyes Data Corporation, 1978)

*Medical Applications*

C. G. Gobelen and F. F. Koblitz, "Biomedical and dental applications of polymers," in *Polymer Science and Technology*, vol. 14 (New York: Plenum, 1981)

M. H. Gutcho, *Microcapsules and Other Capsules* (Noyes Data Corporation, 1979)

P. G. Stecher, ed., *New Dental Materials* (Noyes Data Corporation, 1980)

A. Zaffaroni, "Delivering drugs," *Chemtech* 10:82 (1980)

## Chapter 6

The following works provide the reader with an up-to-date overview of inorganic chemistry and the impact of the modern inorganic-chemicals industry on resource availability, the environment, and society:

H. P. Ableson and L. A. Hammond, eds., *Materials: Renewable and Non-Renewable Resources* (Washington, DC: AAAS, 1976)

*Annuaire Statistique MINEMET* (Paris: 1979)

Commodity Research Unit, *Problems and Perspectives for Raw Materials* (London: 1975)

P. Connolly and R. Perlman, *The Politics of Scarcity: Resources Conflicts in International Relations* (Oxford: Oxford University Press for RITA, 1975)

P. Dasgupta and G. Heal, *Economic Theory and Exhaustible Resources* (Cambridge: Cambridge Handbook Series, 1978)

Economist Intelligence Unit, *Raw Material Prices in the 1980's* (London: 1978)

J. E. Fergusson, *Inorganic Chemistry—The Environment and Society* (New York: Pergamon Press, 1980)

M. L. Jenses and A. M. Bateman, *Economic Mineral Deposits* (New York: Wiley, 1979)

G. Manners, "The future markets for minerals: Some cause of uncertainty," *Resources Policy Conference '78* (Guildford: IPC Science and Technology Press, 1978)

J. L. Mero, *The Mineral Resources of the Sea* (Amsterdam: 1964)

Ministere de l'Indistrie, *Matières Premieres Minerales* (Paris: 1981)

OECD, *L'Industrie des Metaux non Ferreux* (Paris: OECD, 1974)

OECD, *The Chemical Industry* (Paris: OECD, 1979)

B. G. Reuben and M. L. Burstall, *The Chemical Economy* (London: Longmans, 1978)

J. Ridgeway, *Who Owns the Earth* (New York: Macmillan, 1980)

M. Sittig, *Inorganic Chemical Industry—Processed, Toxic Effluents and Pollution Control* (Noyes Data Corporation, 1978)

Stanford Research Institute, *World Minerals Availability 1975–2000* vols. 3 and 5 (Stanford: Stanford Research Institute, 1976)

E. J. Stevenson, *Extractive Metallurgy—Recent Advances* (Noyes Data Corporation, 1977)

R. Thompson, ed., *The Modern Inorganic Chemicals Industry*, special publication no. 31 (London: The Royal Society of Chemistry, 1977)

Union Minière, *Dictionnaire des Metaux non Ferreux* (Brussels: 1972)

References dealing with topics of particular interest are

R. G. Bond and C. P. Straub, eds., *Handbook of Environmental Control*, vols. 1–5 (Boca Raton: CRC Press, 1978)

S. S. Brown and F. W. Sunderman, Jr., eds., *Nickel Toxycology* (New York: Academic Press, 1981)

Commission of the European Communities, *Health Criteria (Exposure/Effect Relationships) for Mercury* (New York: Pergamon Press, 1979)

Department of the Environment, Central Directorate of Environmental Pollution, *Cadmium in Environment and its Significance to Man*, pollution paper no. 17 (London: Her Majesty's Stationery Office, 1980)

E. Di Ferrante, ed., *Trace Metals: Exposure and Health Effects* (New York: Pergamon Press, 1979)

H. L. Needleman, ed., *Low Level Lead Exposure* (New York: Raven Press, 1980)

D. J. Rapport and H. A. Regier, "An ecological approach to environmental information," *Ambio* 9:22 (1980)

J. M. Ratcliffe, *Lead in Man and the Environment* (Chichester: Ellis Horwood, 1981; distributed by Wiley)

M. Sittig, *Toxic Metals—Pollution Control and Worker Protection* (Noyes Data Corporation, 1976)

M. Sittig, *Industrial and Toxic Effects of Industrial Chemicals* (Noyes Data Corporation, 1979)

A. V. Slack and G. A. Hollinden, *Sulfur Dioxide Removal from Waste Gases*, 2nd ed. (Noyes Data Corporation, 1975)

A. C. Stern, ed., *Engineering Control of Air Pollution*, vol. 4 in *Air Pollution* (New York: Academic Press, 1977)

L. H. Yaverbaum, *Nitrogen Oxides Control and Removal—Recent Developments* (Noyes Data Corporation, 1979)

A. Yehaskel, *Industrial Wastewater Clean-up—Recent Developments* (Noyes Data Corporation, 1979)

## Chapter 7

Energy has become a major problem for industrialized and developing countries. The following works give an overview:

P. H. Abelson and A. L. Hammond eds., *Energy II: Use. Conservation, and Supply* (Washington, DC: AAAS, 1978)

Center for Study of the American Experience, *Energy in America: Fifteen Views* (Eastbourne: Holt-Saunders, 1981)

S. D. Christian and J. J. Zuckerman, eds., *Energy and the Chemical Sciences* (New York: Pergamon Press, 1978)

E. Cook, "Charting our energy future," *Chemtech* 11:441 (1981)

R. Eden, M. Posner, E. Crouch, and J. Stanislaw, *Energy Economics: Growth, Resources and Policies* (Cambridge: Cambridge University Press, 1981)

*Education in Chemistry* 15(1) (1978): articles concerned with the energy problem.

J. H. Harker and J. R. Backhurst, *Fuel and Energy* (New York: Academic Press, 1981)

D. C. Ion, *Availability of World Energy Resources* (London: Graham and Trotman, 1975)

H. H. Landberg, ed., *Energy: The Next Twenty Years* (Cambridge, MA: Ballinger, 1979)

OECD, *Bilans Energetiques des Pays de l'O.C.D.E. 1974–1978* (Paris: OECD, 1980)

OECD, *Statistiques de l'Energie 1974–1978* (Paris: OECD, 1980)

W. Sassin, "Urbanization and the global energy problem," in *Factors Influencing Urban Design—A Systems Approach*, P. Laconte, J. Gibson, and A. Rapport, eds. (Amsterdam: Sijthoff and Noordhoff, 1980)

K. R. Stunkel, ed., *National Energy Profiles* (Eastbourne: Holt-Saunders, 1981)

Twentieth Century Fund, Inc., *Providing for Energy: Report of the Twentieth Century Fund Task Force on the United States Energy Policy* (New York: McGraw-Hill, 1977)

Workshop on Alternative Strategies, *Energy Global Prospects 1985–2000* (New York: McGraw-Hill, 1977

For studies on specific topics, see

The Conference Board, *Energy Consumption in Manufacturing* (Cambridge, MA: Ballinger, 1974)

R. C. Fluck and C. Direlle Baird, *Agricultural Energetics* (Westport, CT: AVI, 1980)

M. G. Green, *Eating-Oil Energy Use in Food Production* (Boulder, CO: Westview Press, 1978)

R. C. Loehr, ed., *Food Fertilizer and Agricultural Residues* (Ann Arbor, MI: Ann Arbor Science, 1977)

R. C. Loehr, *Food, Fertilizer and Agricultural Residues* (Ann Arbor, MI: Ann Arbor Science, 1977)

A. Makhajani, *Energy and Agriculture in the Third World* (Cambridge, MA: Ballinger, 1978)

L. B. McGown and J. Mockris, *How to Obtain Abundant Clean Energy* (New York: Plenum, 1980)

D. Pimentel, ed., *Energy Utilization in Agriculture* (Boca Raton: CRC Press, 1980)

D. Pimentel, ed., *Handbook of Energy Utilization in Agriculture* (Boca Raton: CRC Press, 1980)

D. Pimentel and M. Pimentel, *Food, Energy and Society* (London: E. Arnold, 1979)

M. Sittig, *Practical Techniques for Savings Energy in the Chemical Petroleum and Metals Industries* (Noyes Data Corporation, 1977)

M. L. Sittig, ed., *The Utilization and Recycle of Agricultural Wastes and Residues* (Boca Raton: CRC Press, 1980)

J. S. Steinhart and C. E. Steinhart, "Energy use in the U.S. food system," in *Energy II: Use, Conservation and Supply*, P. H. Abelson and A. L. Hammond, eds. (Washington, DC: AAAS, 1978)

Additional information on environmental problems and conventional and so-called alternative energy sources can be found in the following current works:

*Energy: Safety and Environment*

H. Ashley, R. L. Rudman, and C. Whipple, *Energy and Environment: A Risk-Benefit Approach* (New York: Pergamon Press, 1976)

W. P. Beaton, *Energy Forecasting for Planners: Transportation Models* (Eastbourne: Holt-Saunders, 1981)

R. K. Coyne and R. J. Clark, *Environmental Assessment and Design* (Eastbourne: Holt-Saunders, 1981)

Health and Safety Commission, *The Hazards of Conventional Sources of Energy* (London: Her Majesty's Stationery Office, 1978)

L. B. McGown and J. O. Mockris, *How to Obtain Clean Energy* (New York: Plenum, 1980)

N. C. Rasmussen, *Reactor Safety Study—An Assessment of Accident Risks in U.S. Commercial Nuclear Power Plants* (United States Nuclear Regulatory Commission, WASH-1400-NUREG 75/104, 1975)

F. A. Robinson, ed., *Environmental Effects of Utilizing More Coal* (London: The Royal Society of Chemistry, 1980)

A. Stern, ed., "Engineering control of air pollution", vol. 4 in *Air Pollution* (New York: Academic Press, 1977)

WHO, *Health Implications of Nuclear Power Production* (Copenhagen: WHO Regional Publication European Series no. 3, 1978)

*Energy Sources: Hydrocarbon Fuel*

D. K. Rider, *Energy: Hydrocarbon Fuels and Chemical Resources* (New York: Wiley, 1981)

*Energy Sources: Coal*

R. T. Ellington, ed., *Liquid Fuels from Coal* (New York: Academic Press, 1977)

R. Noyes, ed., *Coal Resources, Characteristics and Ownership in the U.S.A.* (Noyes Data Corporation, 1977)

A. H. Pelofsky, ed., *Coal Conversion Technology Problems and Solutions* (Washington, DC: American Chemical Society, 1980)

G. J. Pitt and G. R. Millward, eds., *Coal and Modern Coal Processing: An Introduction* (New York: Academic Press, 1979)

*Energy Sources: Fission and Fusion*

Commission of European Communities, *Nuclear and Non-Nuclear Risk—an Exercise in Comparability* (EUR/6417/EN, 1980)

R. Curtis and E. Hogan, with S. Horowitz, *Nuclear Lessons: An Examination of Nuclear Power's Safety, Economic, and Political Record* (Harrisburg, PA: Stackpole Books, 1980)

J. J. Dunderstadt and C. Kikucki, *Nuclear Power: Technology on Trial* (Ann Arbor, MI: University of Michigan Press, 1979)

G. Greenhalg, *The Necessity of Nuclear Power* (London: Grahan and Trotman, 1980)

R. Jungk, *The New Tyranny: How Nuclear Power Enslaves Us* (New York: Grosset and Dunlop, 1979)

*Developing Energy: General*

*Ambio* 10 (1981), articles on alternative energy sources.

A. B. Lovins, *Soft Energy Paths* (London: Penguin Books, 1977)

*Developing Energy: Solar Energy*

C. C. Black and A. Mitsui, eds., *Fundamental Principles,* vol. 1 in *CRC Handbook of Biosolar Resources*, O. R. Zaborsky, series ed., (Boca Raton: CRC Press, 1981)

J. S. Connolly, ed., *Photochemical Conversion and Storage of Solar Energy* (New York: Academic Press, 1981)

F. Kreith and R. E. West, *Economics of Solar Energy and Conservation Systems* (Boca Raton: CRC Press, 1980)

W. D. Metz and A. L. Hammond, *Solar Energy in America* (Washington, DC: AAAS, 1978)

A. Mitsui, S. Miyachi, A. San Pietro, and S. Tamura, eds., *Biological Solar Energy Conversion* (New York: Academic Press, 1977)

National Energy Research Programme, *Heat From Solar Energy* (The Hague: Government Printing Office, 1978)

J. K. Paul, ed., *Passive Solar Energy Design and Materials* (Noyes Data Corporation, 1979)

A. Rose, "Solar energy: A global view: I," *Chemtech* 11:566 (1981)

*Developing Energy: Biomass and Wastes*

E. J. Da Silva, "Biogas: Fuel of the future?" *Ambio* 9:2 (1980)

D. J. De Renzo, ed., *European Technology for Obtaining Energy From Solid Waste* (Noyes Data Corporation, 1978)

F. A. Domino, ed., *Energy from Solid Waste* (Noyes Data Corporation, 1979)

I. S. Goldstein, ed., *Organic Chemicals From Biomass* (Boca Raton: CRC Press, 1981)

J. L. Jones and S. B. Radding, eds., *Thermal Conversion of Solid Wastes and Biomass* (Washington, DC: American Chemical Society, 1980)

K. Z. Sarkanen and D. A. Tillman, eds., *Progress in Biomass Conversion*, vol. 1 (New York: Academic Press, 1979)

D. A. Stafford, D. L. Hawkes, and H. R. Norton, *Methane Production from Waste Organic Matter* (Boca Raton: CRC Press, 1980)

A. van Buren, "Biogas beyond China: First international training program for developing countries," *Ambio* 9:10 (1980)

D. L. Wise, ed., *Fuel Gas Production from Biomass* vols. 1–2 (Boca Raton: CRC Press, 1981)

*Developing Energy: Wood Energy Conversion*

R. C. Myerly, M. D. Nicholson, R. Katzen, and J. M. Taylor, "The forest refinery," *Chemtech* 11:186 (1981)

D. A. Tillman, *Woods as an Energy Resource* (New York: Academic Press, 1977)

D. A. Tillman, K. V. Sarkanen, and L. L. Anderson, eds., *Fuels as Energy from Renewable Resources* (New York: Academic Press, 1977)

C. R. Wilkes, ed., *Cellulose as a Chemical and Energy Resource* (New York: Wiley, 1975)

*Developing Energy: Geothermal Energy*

M. C. Collie, ed., *Geothermal Energy—Recent Developments* (Noyes Data Corporation, 1978)

J. Goguel, S. P. Clark, and A. C. Rite, *Geothermics* (New York: McGraw-Hill, 1975)

*Developing Energy: Winds and Waves*

D. J. De Renzo, ed., *Wind Power—Recent Developments* (Noyes Data Corporation, 1979)

M. E. McCormick. *Ocean Wave Energy Conversion* (New York: Wiley, 1981)

**Chapter 8**

*The "Good Old Days"*

T. Barker, *The Long March of Everyman—1750–1960* (London: Penguin Books, 1978)

O. L. Bettmann, *The Good Old Days: They Were Terrible* (New York: Random House, 1974)

C. M. Cipolla, *The Economic History of World Population* (New York: Penguin Books, 1962)

C. M. Cipolla, *Storia Economica dell'Europa Pre-Industriale* (Bologna: Il Mulino, 1975)

*Smoking*

H. S. Diehl, *Tobacco and Your Health* (New York: McGraw-Hill, 1969)

H. V. Gelboin and P. O. P. Ts'o, "Polycyclic hydrocarbons and cancer: Environment, chemistry, molecular and cell biology," in *Environment, Chemistry and Metabolism*, vol. 1 (New York: Academic Press, 1978)

R. M. Greenhalg, ed., *Smoking and Arterial Disease* (London: Pitman Books, 1981)

*Alcohol*

J. Fort, *Alcohol: Our Biggest Drug Problem* (New York: McGraw-Hill, 1973)

O. Josseau Kalant, ed., *Alcohol and Drug Problems in Women* (New York: Plenum, 1980)

J. S. Madden, R. Walker, and W. H. Kenyon, eds., *Aspects of Alcoholism and Drug Dependence* (London: Pitman Books, 1980)

M. A. Schuckit, *Drug and Alcohol Abuse: A Clinical Guide to Diagnosis and Treatment* (New York: Plenum, 1979)

*Weapons*

Center for the Study of Armament and Disarmament, "Chemical/biological warfare: A selected bibliography," *Political Issues* 6(2) (1979)

Committee on Toxicology, National Research Council, *Review of U.S. Air Force Protocol: Epidemiological Investigation of Health Effects in Air Force Personnel Following Exposure to Herbicide Orange* (Washington, DC: National Academy of Sciences, 1980)

"Dioxin traces found in soldiers exposed to defoliant," *Nature* 282:772 (1979)

L. R. Ember, "Chemical weapons: Build up or disarm?" *Chem. and Eng. News* 58:22 (1980)

J. Brown Frederic, *Chemical Warfare: A Study in Restraints* (Princeton: Princeton University Press, 1968)

M. Meselson, ed., *Chemical Weapons and Chemical Arms Control* (Carnegie Endowment for International Peace, 1978)

J. P. Robinson, "The effects of weapons on ecosystems," in *UNEP Studies*, vol. 1 (New York: Pergamon Press, 1979)

SIPRI (Stockholm International Peace Research Institute), *The Problem of Chemical and Biological Warfare* (Stockholm: 1971–1975)

SIPRI (Stockholm International Peace Research Institute), *Chemical Weapons: Destruction and Conversion* (Crane, Russak, 1980)

SIPRI (Stockholm International Peace Research Institute), "World armaments and disarmaments," *SIPRI Yearbook 1980* (London: Taylor and Francis, 1981)

## Chapter 9

The balance of risks and balances is weighed in

S. C. Black, F. Niehaus, and D. Simpson, *How Safe is "Too" Safe* (Laxenburg, Austria: International Institute for Applied Systems Analysis, WP-79-68, 1979)

B. L. Cohen and J. Sing Lee, "A catalog of risks," *Health Physics* 36:6 (1979)

J. Conrad, ed., *Society, Technology and Risk Assessment* (New York: Academic Press, 1980)

B. Crickmer, "Regulation: How much is enough?" *Nation's Business* 26–33 (March 1981)

B. Fischhoff, P. Slovic, S. Lichtenstein, S. Read, and B. Coombs, "How safe is safe enough? A psychometric study of attitudes toward technological risks and benefits," *Policy Sciences* 9:127 (1978)

R. F. Griffiths, ed., *Dealing with Risk: The Planning Management and Acceptability of Technological Risk* (Halstead: Manchester University Press, 1981)

R. W. Kates, *Risk Assessment of Environmental Hazard*, no. 8 in SCOPE Series (New York: Wiley, 1981)

W. W. Lowrance, *On Acceptable Risk* (Los Altos, CA: William Haufmann, 1976)

W. D. Rowe, *An Anatomy of Risk* (New York: Wiley-Interscience, 1977)

C. Starr, "Social benefit versus technological risk," *Science* 165:1232 (1969)

The following readings touch on trends in the regulation of chemicals:

C. M. Binford, C. Fleming, and Z. A. Prust, *Loss Control in OSHA Era* (New York: McGraw-Hill, 1975)

B. E. Butterworth, *Strategies for Short-Term Testing for Mutagens/Carcinogens* (Boca Raton: CRC Press, 1979)

N. P. Cheremisinoff, P. N. Cheremisinoff, F. Ellerbusch, and A. J. Perna, *Industrial and Hazardous* (New York: Wiley, 1979)

H. E. Christensen and T. T. Luginbyhl, eds., *Registry of Toxic Effects of Chemical Substances* (Washington, DC: US Dept. of Health, Education and Welfare, 1975)

Commission du Codex Alimentarius, *Guide Concernant les Limites Maximales Codex pour les Residus de Pesticides* (FAO-OMS, CAC/PR 1, 1978)

Commission of the European Communities, *Criteria (Dose/Effect Relationships) for Humans on Organochlorine Compounds* (New York: Pergamon Press, 1979)

Commission of the European Communities, *Health Criteria (Exposure/Effect Realtionships) for Mercury* (New York: Pergamon Press, 1979)

F. Coulston, ed., *Regulation Aspects of Carcinogens and Food Additives: The Delaney Clause* (New York: Academic Press, 1980)

G. S. Dominguez, ed., *Guidebook: Toxic Substances Control Act* (Boca Raton, CRC Press, 1977)

F. H. Gross and W. H. W. Inman, *Drug Monitoring* (New York: Academic Press, 1977)

Y. H. Hui, *United States Food Laws, Regulations and Standards* (New York: Wiley, 1979)

F. W. Mackison, R. Scott Stricoff, and L. J. Partridge, eds., *Occupational Health Guidelines for Chemical Hazards* (Washington, DC: US Government Printing Office, 1981)

OECD, *OECD and the Environment* (Paris: OECD, 1979)

OECD, *OECD Guidelines for Testing of Chemicals* (Paris: OECD, 1981)

M. E. Simon, "What regulation costs," *Chemtech* 11:100 (1981)

Toxic Substances Strategy Committee, *Toxic Chemicals and Public Protection* (Washington, DC: US Government Printing Office, 1980)

I. Walter and E. J. Ugelow, "Environmental policies in developing countries," *Ambio* 8:102 (1979)

M. Weidenbaum, "America's greatest growth industry," *Chemtech* 10:292 (1980)

# Index